Transforming New Orleans and Its Environs

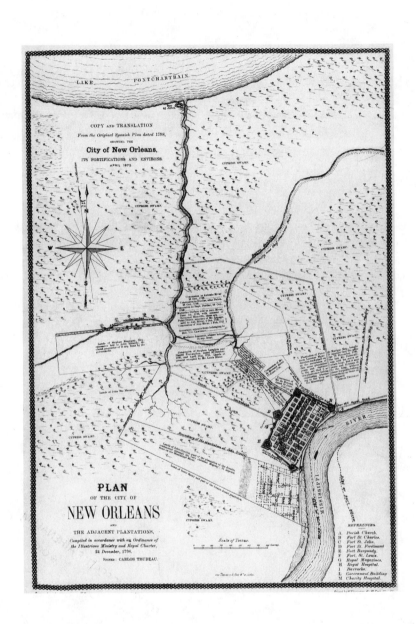

COPY AND TRANSLATION
From the Original Spanish Plan dated 1798,
SHOWING THE
City of New Orleans,
ITS FORTIFICATIONS AND ENVIRONS.
APRIL 1875

LAKE PONTCHARTRAIN.

CYPRESS SWAMP.

PLAN
OF THE CITY OF
NEW ORLEANS
AND
THE ADJACENT PLANTATIONS,
Compiled in accordance with an Ordinance of
the Illustrious Ministry and Royal Charter,
24 December, 1798.
SIGNED: CARLOS TRUDEAU.

Scale of Toesas.

REFERENCES.
A   Parish Church.
B   Fort St. Charles.
C   Fort St. John.
D   Fort St. Ferdinand
E   Fort Burgundy.
F   Fort. St. Louis.
G   Royal Magazines.
H   Royal Hospital.
I   Barracks.
L   Government Building
M   Charity Hospital.

Transforming
New Orleans
and
Its Environs

CENTURIES OF CHANGE

Craig E. Colten, editor

University of Pittsburgh Press

Library of Congress Cataloging-in-Publication Data
Transforming New Orleans and its environs : centuries of change / Craig
E. Colten, editor.
    p. cm.
Includes bibliographical references and index.
    ISBN 0-8229-4134-1 (acid-free paper) — ISBN 0-8229-5740-X (pbk. :
acid-free paper)
    1. Nature—Effect of human beings on—Louisiana—New Orleans Region.
2. Human ecology—Louisiana—New Orleans Region—History. I. Colten,
Craig E.
    GF504.L8 T73 2000
    304.2'8'0976335—dc21

                                                              00-011648

Contents

List of Figures

Chapter 9

Chapter 10

Chapter 11

Chapter 12

Acknowledgments

This volume represents the efforts of several individuals and institutions. First I must point out the important role played by Edwin Lyon of the U.S. Army Corps of Engineers, New Orleans District. Before the inception of this volume, I turned to Edwin Lyon for guidance on a new research project centered on New Orleans and its environmental history. Basic assistance turned into something much more extensive. Through his connections with and involvement in Tulane University's History Department, we developed a plan to assemble a group of scholars to discuss environmental change on the lower Mississippi. Edwin Lyon and the Corps deserve credit for pulling together the players who ultimately put on the first Randall L. Gibson Symposium, titled "Centuries of Change: Human Transformation of the Lower River," from whence this volume derives. He also identified most of the presenters and secured the participation of several.

Terrance Fitzmorris and Tulane University were principal players in the events leading up to this volume as well. Fitzmorris, the newly formed Center for the Study of New Orleans and the Mississippi River, and Tulane's University College provided support and served as hosts for the conference in October 1998. Tulane scholars were also key contributors to this volume.

My academic home at the time, Southwest Texas State University, also provided essential support. The Lovell Center for Environmental Geography and Hazards Research and the Department of Geography offered generous financial assistance for the participants and the conference. In addition, the center allowed Wendy Bigler to assist with final manuscript preparation. She deserves my deepest gratitude for her unglamorous but vital editorial contribution. Shannon Crum deftly and efficiently prepared the map to accompany the introduction and also has my gratitude for her contribution.

The authors, of course, are the most important ingredient of any collection of essays. Most participated in the Centuries of Change conference, which offered a lively forum for discussing environmental change on the lower river. Others submitted papers that helped round out the volume. All made a valued scholarly contribution.

I would also like to acknowledge and thank Andrew Hurley of the University of Missouri–St. Louis for offering such insightful comments on a draft of the manuscript. His suggestions and that of a second anonymous reader greatly improved this work. Neils Aaboe of the University of Pittsburgh Press diligently guided this project along in a timely and professional manner. As editor, I wish to thank these individuals and institutions who contributed to this effort.

Transforming New Orleans and Its Environs

I

Craig E. Colten

Introduction

Transforming the Lower Mississippi River Valley

T HE MAJORITY OF Americans live in cities today. Wrapped in an urban infrastructure, we easily forget that cities occupy places carved out of the natural environment and that city building transforms the local setting. While pavement and structures obscure local geology and hydrology, natural systems continue to operate. Making or remaking a place to be habitable is a long-term process that involves many generations. This volume considers one American urban area, New Orleans and the lower Mississippi River valley, to expose the lineaments of environmental systems, how they continue to function within a metropolitan region, and how human actions over time have altered the place that is occupied by today's city.

New Orleans truly is shaped by its environment, but at the same time its builders transformed the natural landscape. Its streets radiate out from a giant bend in the river, early settlement clustered on the better-drained natural levees, and it owes its urban stature to its gateway location for the Mississippi River valley. The river as a thoroughfare presented commercial opportunities, but it also posed threats, particularly in the form of floods. Its backwater swamps provided cypress for building material and habitat for disease-bearing vectors. Its soggy terrain challenged engineers seeking solid footing for railroads, highways, and buildings. To make the site suitable for a city, Indian and European settlers had to reshape the topography to suit the demands of urban life. In doing so they began the process of modifying natural systems on a local level. These transformations, in most cases, produced improved living conditions in the short term, but sometimes resulted in negative outcomes for dwellers beyond the city limits.

Although in many respects New Orleans is peculiar, the focus on environmental transformation of this quirky city has broader import. Settlement on the wetland ridges by native people was part of a larger expansion of indigenous people finding ways to make corners of the countryside habitable. French colonizers played a role in the vast transatlantic extension of European culture. Engineers, both private and government, undertook their duties within the limits of contemporary knowledge of hydrology when they designed flood-control systems. Industries and residents of the lower valley used the river according to their capabilities and the larger environmental and legal framework of the day. Human transformations occurred within a temporal and geographic context that was larger than the territory under discussion. So while this volume seeks to add crisp detail to the story of environmental change in and around New Orleans, it also will contribute to the more general discussion of environmental history of urban areas.[1]

Geographer Carl Sauer spoke of the "natural landscape" as the "datum line from which [human-induced] change is measured."[2] The social philosopher Karl Marx used the term *first nature* to refer to the environment unaltered by human labor. While these terms are deeply imbued with ideological and methodological significance, they basically refer to the same thing: portions of the earth that have been materially unchanged by human actions. To seek out such untouched territory assumes that there are quarters without the imprint of society. But as the recent volume, *Human Impact on the Earth,* informs us, there is no remote corner of the earth that has not been transformed to some degree.[3] Nonetheless, natural processes operate despite human intervention, and in some cases they modify human creations. New Orleans and the lower Mississippi River valley is one such area where human agency has been active for centuries, frequently struggling with natural forces and always leaving at least a transitory imprint.

One way to approach environmental transformation stimulated by the lower river's urban complex would be to follow the impressive effort of William Cronon. His *Nature's Metropolis* examines the impact of Chicago on its sprawling hinterland, how the market modified nature.[4] Certainly one could look at the extensive territory tributary to New Orleans and consider the effect of trade opportunities on the forests and fields of the upper valley. This work, however, will take a much more localized approach and consider the local transformations—to physical landforms, fish life, humans, and the river itself. These changes were fundamental to the functioning of the city and its neighboring environs and were basic to the growth of an important entrepôt. Likewise, as New Orleans's economic importance rose, the greater the resources expended on environmental manipulations. In order to protect the burgeoning metropolis, urban dwellers had to take massive transformative steps. They selected courses of action that protected their houses and businesses and sometimes endangered neighboring smaller communities and rural dwellers. Thus, the urban center of the area had a tremendous influence on policy and practice as they relate to environmental change even beyond its boundaries.

This volume presents essays that explore the transformative process. Some scholars refer to human modifications as *degradation*.[5] While this term is appropriate in many cases, it does not fit all human modification of the environment, especially when considering long-term habitation or industrial impacts. *Transformation* fits better because it acknowledges that not all actions reduce the value of land or other environmental resources. Further, it allows us to place modifications in their historical context as actions that may have reflected laudable objectives in the past, that may or may not have produced deleterious results over time. The scale and extent of transformation obviously vary depending on the size of the population, the political power to mobilize groups of people, and the technological capabilities of those at work. This volume will consider both small alterations, locally and socially significant, along with extensive modifications with regional and national import. It will also look at both direct and indirect consequences of human actions. Throughout, the unifying theme will be that human action has inscribed a legacy on the Mississippi delta region wherever individuals hunted, gathered, built on, tilled, processed, or extracted objects of nature. Human transformation aimed at providing a habitable environment is certainly not unique to this setting, but it was vital and the results of past actions remain present in current efforts to deal with those historical actions. Also, to understand the relationship between human actions, the environment, and current conditions demands a historical perspective. For this reason the contributors trace actions over the past several centuries and present the current landscape as a composite from multiple generations of humans.

The focal point for this investigation will be the lowermost Mississippi River valley (see fig. 1.1). Not the official Corps of Engineers' lower river (from Cairo to the mouth), but a much smaller territory—from approximately where the river leaves the bluff line near Baton Rouge to the tip of the bird's foot delta. This territory is selected because there has been a long tradition of studies of the deltaic land as a unit of observation, that relatively new land built by the meanderings of the river.[6] Within this territory, recognizable native American cultures lived. It was the focus of early French settlement. In reference to the lower valley, geographer and long-time lower river resident Fred Kniffen claimed that "two primary molding forces set the course of development": the river and Euro-American addiction to agriculture.[7] Both, of course, required extensive modification for the current pattern of occupance to unfold. The relationship of settlers and developers to the river, along with their desire to fashion an agricultural domain on the floodplain, are central to understanding this region. More recently, the concentration of industry has earned the New Orleans–Baton Rouge corridor the designation of "cancer alley." Noted as a significant petrochemical complex, the linear cluster of plants selected their sites for access to river water and transportation. Thus, even as land use has undergone a transition from agriculture to industry, there has been a unity to the river region that provides a logical basis for examining environmental change here.

Peirce Lewis characterized New Orleans, the dominant urban settlement of the lower river valley, as an "impossible but inevitable city."[8] While overstating both points, Lewis highlights two key geographic concepts that apply to the lower valley. A major entrepôt was essential for a colonial power to claim the vast drainage basin of the Mississippi, and the low-lying, flood-prone site selected by the French destined the City of New Orleans to a perpetual struggle with riparian forces. Frequent floods, a wetland site, and the attendant pests and diseases have vexed settlers of the lower river since the city's inception. Native Americans preceded the French in occupying the shifting deltas of the lower Mississippi and, like current society, they had to modify the marsh to make it habitable. Over time, the fundamental purposes of modification have remained similar, although impacts have become more extensive. Inhabitants directed the first major transformation efforts toward reducing the impact of high water—whether high water table

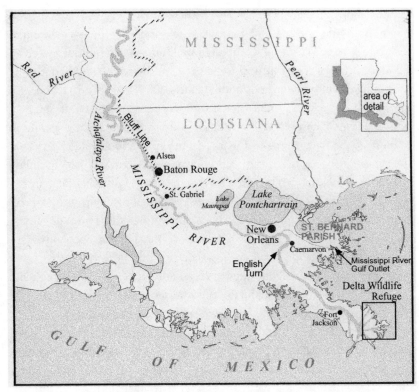

Figure 1.1. Lower Mississippi River region.

or floodwater. Middens in the marsh recount the filling activities of prehistoric societies. Levees and other flood-control structures are the modern counterparts. A second transformation focused on subsistence efforts—whether agricultural or industrial. Farming the natural levees and the eventual replacement of pastures and row crops with cities and chemical plants has produced a series of impacts felt in the lower river valley. With a greater concentration of people and infrastructure, the efforts to manipulate the river and floodplain have become more pronounced. Flood-protection levees now ring the city, expensive pollution-control devices inhibit toxic releases, and navigation channels cut through the marsh alter shipping routes and natural habitat. The chapters in this volume present the increasingly complex and extensive impacts of an urban and industrial society.

To make sense of the human efforts to rework the environment, this volume is organized into four sections. The first, "Transformation before

Urbanization," considers native and French encounters with the environment. Both the Indians and European colonizers brought rudimentary technology to the scene, and, although their efforts were minor by today's standards, they were of utmost importance in their time and hence in determining the ultimate urbanization of the place. The second section, "Environment in Service of the City," considers three elements of struggles with the river. These nineteenth-century efforts illustrate the expenditure to manage the river, to make it work as a transportation route in service to the city, and to deny its unwanted floodwaters access to city streets and the farmlands that sustained the growing urban center. "Growing Demands of the City," the third section, traces the displacement of environmental consequences resulting from efforts to prevent natural hazards from destroying the city. New Orleans by the 1920s was the dominant social, economic, and political force in the lower valley. To save itself from the flood of 1927, it was willing to sacrifice a neighboring parish and the wildlife its residents relied on. Again in the 1960s, to maintain its economic position New Orleans's interests were willing to sacrifice the wetlands of St. Bernard Parish. Finally, "Response to Environmental Change" examines the social and environmental effects of using the environmental as a sink for industrial wastes. Drinking water, neighbors of the industrial plants, and aquatic life all endured impacts. While not a complete or comprehensive account of environmental transformations in the New Orleans area, this work offers perspectives on critical aspects of human encounters with natural processes.

Transformation before Urbanization

Cities stand as the ultimate expression of human transformation of the natural environment. Streets conquer topography, and structures obscure the natural lay of the land and displace vegetation. Drainage and sewerage systems reconfigure the local hydrology. Urban civilization was not unknown in either the Americas or in the Mississippi valley when Europeans arrived. Cahokia, near present-day St. Louis, was perhaps the greatest example north of the Rio Grande; and it remained the largest city in the Mississippi valley even after the founding of New Orleans. But urbanization is not the only modification wrought by human society. In the New World, there had been extensive manipulation of nature beyond the major settlements. While not as concentrated, prehistoric activity left a significant imprint— particularly for the purposes of the prehistoric peoples.

Likewise, colonial explorers and settlers modified the environment they encountered, perhaps tentatively at first, as they acquired knowledge about plants and animals from native people, and then more extensively as their numbers grew and their technology evolved. The French established New Orleans in 1718, declaring the tenuous settlement a city from the outset. Its imprint was modest to say the least: a grid pattern of streets encompassing some forty blocks and a dirt palisade mounded up around its perimeter. From the city's founding, residents have grappled with myriad environmental problems both within the city limits and in the surrounding territory. What the French sought in Louisiana was not just a city, but a productive hinterland as well. Their efforts to establish an agricultural colony to support an urban center led to efforts beyond the bastide on the Mississippi, to a larger territory engulfed by the present city. This increased the scale and extent of intentional modifications.

Historian William Cronon asserts that, along with language, a basic trait of human groups is their conscious drive to change their environments. He also argues that "one of the best measures of a culture's ecological stability may well be how successfully its environmental changes maintain its ability to reproduce itself."[1] This section will deal with the ecological relations of two groups: the native Indian societies and the colonizing French. The former represented a series of groups who lived in the wetlands around New Orleans since the time when the land had become habitable. By their presence and their selection of foodstuffs, they began to alter the terrain with midden heaps. Through their choice of settlement sites they created a prehistory of human habitation that was critical in subsequent French colonization. Their imprint was gradual and perhaps imperceptible to colonists, but, as William Denevan has suggested, "even mild impacts and slow changes are cumulative, and the long-term effects are dramatic."[2] Tristram Kidder makes this point in his essay on the impacts of prehistoric occupants of the region.

Christopher Morris examines the efforts of the French to become sufficiently familiar with a foreign environment so that they could first survive and eventually thrive. Although initially under the impression that it would be easy to subdue the land, the French encountered failure, followed by experimentation that led to adaptations in their efforts to inhabit the delta swamplands. As Morris explains, they found it easier to live in the environment than to control it, though they worked toward the latter.

Together these two essays demonstrate that human-induced environmental change began centuries ago and was significant to the inhabitants of the site of New Orleans then as well as to its current residents. Native geosophy (informal geographic knowledge) prefaced French selection of the site and their adoption of many agricultural crops and techniques. The ability of both societies to find a way to occupy the site of present-day New Orleans created the possibility that subsequent societies would continue to use the location. Modification has been continual and reflects knowledge of environmental conditions, adaptation to changing conditions, and a perpetual drive to reshape nature to suit the occupants of a place. Given the role of native and French settlers, Kidder and Morris both argue that the site was not the inevitable location of a major city, as suggested by Peirce Lewis.[3]

2   *Tristram R. Kidder*

Making the City Inevitable

Native Americans and the Geography of New Orleans

THE GEOGRAPHER Peirce Lewis has called New Orleans an "inevitable city," asserting that its colonization and subsequent development were influenced by the particular geography of the Mississippi River and the deltaic plain. Discussions of New Orleans as an inevitable city suggest that European colonists chose the site for strategic reasons without consideration of the role existing Native American habitation played in the selection of the site for settlement. Such thinking reflects one of two popular explanations for how humans interact with the physical environment. The first can be identified with the "Ecologically Noble Savage" and the other *"Homo devastans."*[1] The Ecologically Noble Savage is Rousseauesque in his qualities. Human nature, in this doctrine, is benign. The Ecologically Noble Savage does not work actively to diminish biodiversity. In some instances, the Noble Savage may promote biodiversity. In all cases, this archetype demonstrates ecological sensitivity, wisdom, and dignity. Furthermore, the Ecologically Noble Savage represents the sum of indigenous values and knowledge; these values are more discerning than any promoted by Western scientific thinking and are manifest by the neutrality of the indigenous effect on the environment.[2]

Homo devastans, however, is the archetype of ecological villainy. This doctrine of human nature suggests that it is in the human genome to pollute, despoil, and waste natural resources.[3] Those who hold this belief, which could be argued to be the prevailing worldview of many environmentalists, argue that humans everywhere lower biodiversity by polluting their local environments.[4] It is not in our biological or cultural temperament to act as stewards or managers of nonhuman natural resources. Although not explicitly considered in most instances, it is apparent that these views of human nature are thought to differentiate indigenous from Western value systems.

Historical accounts frequently emphasize these polar opposites to explain how environments have been transformed over time. Many see premodern cultures as environmentally and ecologically transitory and neutral. Their effect on the land and on resources is minimal or not visible. Their ability to transform landscapes is at such a reduced scale that they leave no lasting consequence. This perceived impermanence is especially evident in North America. The notion that premodern peoples did not (or could not) effectively or significantly alter their environment comes together in two separate strands of popular myth-history. Some consider the modern native inhabitants of North America as the first ecologists and environmentalists. A heroic myth has emerged in which the Native Americans lived in "harmony" with the land, neither taking more than they needed nor despoiling or altering the land.[5] The prevalence of this picture in popular media and Native American myth-history is significant and shapes the present-day discussion of the fate of America's remaining "wilderness." Similarly, however, the myth-history of the European settlers of North America worked also to minimize the role premodern Native Americans had in affecting their environment. This notion dates back to the first colonists and is perpetuated even today.

> The grand invented tradition of American nature as a whole is the pristine wilderness, a succession of imagined environments which have been conceived as far more difficult for settlers to conquer than they were in reality. . . . The ignoble savage . . . was invented to justify dispossession . . . and to prove that the Indian had no part in transforming America from Wilderness to Garden.[6]

In a curious irony, the myth-histories of groups at vastly different ends of the philosophical spectrum coincided to perpetuate two myths. On the one hand was the notion of the precontact Forest Primeval, untouched and unaltered by the natives and heroically conquered by the European colonists.

On the other hand was the myth of the passive "ecologically invisible" Indian,[7] the prototype for today's environmentalist. The combined myths resulted in a popular image of

> Indians who lived . . . in harmony with nature, making no irremediable changes in the environment, and handing over to the Europeans a virgin land. Whether denigrated as ignoble savages or idealized as native Americans living in perfect equilibrium and harmony with the environment, the Indians are given no credit for opening up the Eastern Woodlands, for creating much of America's grassland, and for transforming hardwoods to piney woods.[8]

Despite ample evidence to the contrary, indigenous peoples rarely are recognized as active, determinant agents of environmental change or transformation.[9] Usually, historical narratives present Native Americans as bystanders in the great colonial effort to recast the landscape in a suitable fashion for their continental tastes.

How might we reconcile both our language and, more critically, our thinking about human relations with the environment? I wish to set the issue of native roles in shaping the environment of New Orleans against the framework of historical ecology. Historical ecology does not regard nature and culture as a dichotomy; rather, it requires that we see this relationship as a dialogue.[10] In this context, we can reconsider whether New Orleans is an inevitable city.[11] Moreover, if it is, why? Lewis has argued that New Orleans came about because the geography of the Mississippi alluvial plain dictated its presence. The unique location, adjacent to the river with connection via Lake Pontchartrain to the sea, made it a strategically logical, seemingly natural site to place the newly founded colonial capital. In identifying the reasoning for the siting of New Orleans, Lewis and others have relegated native peoples to a negligible or marginal position in the story of how this city evolved. In fact, these native peoples played a vital part in shaping the local ecology of what would become New Orleans, providing added incentive for colonizing this specific location. In addition, it was the native inhabitants of the lower Mississippi valley who demonstrated the strategic location of the New Orleans area. The geographic knowledge of the existing inhabitants of the region was transmitted to the colonists who were quick to absorb the lesson of their native tutors.

Two aspects of the modern landscape are relevant to a discussion of the role of Native Americans in the founding of New Orleans. First, I look at the marsh as an exemplar of historical ecology. Then I explore the role Bayou St. John plays in shaping the city. The result is modern New Orleans, a city influenced by native peoples of south Louisiana who acted as active agents that transformed the land and the colonial experience.

THE MARSH

The marshes of southern Louisiana are frequently cited as examples of natural landscapes. Several chapters in this volume discuss how these marshes were "transformed" in historic and recent times, and how these changes impact the historical development of New Orleans and its surrounding regions. We might question, however, if these marshes represent an example of the "overwhelming" nature that natives encountered and then passed unchanged on to the colonial transformers and manipulators. Instead, I would urge recognition that the marshes of south Louisiana are not now and have never been natural environments if we define nature to exclude the effects of humans as part of nature.[12]

The marshes of south Louisiana are monotonous expanses of grasses interspersed with occasional islands of trees and shrubs. This characterization of the marshes is not intended to be an aesthetic judgment but an ecological one. In the Mississippi River delta, elevation above sea level most strongly influences marsh vegetation.[13] Elevation is, in turn, an element related to larger patterns of sea-level dynamics, subsidence, and delta growth.[14] Marshes in this region are very rich in plants, animals, and insects. Resource distribution in the marsh is not, however, homogeneous.[15] Human-modified contexts—oil canals, levees, and archaeological sites—are important elements of marsh ecology today and influence the kinds and the location of vegetation and other biota within the deltaic plain.[16] Archaeological sites are, in fact, ecological islands of diversity in an otherwise monotonous sea of grasses.[17]

Humans have been exploiting the marshes since their inception as "natural" environments. The earliest sites in the region have been dated to ca. 2000 B.C. Native Americans continuously occupied the region to the end of the eighteenth century. Up to roughly A.D. 1200 most inhabitants subsisted by hunting, gathering, and especially fishing. By the thirteenth century agriculture became more important but never completely supplanted the use of wild foods. Numerous archaeological sites in the area show that native inhabitants were well adapted to the local environment. They were generally healthy and lived in small settlements that were occupied for a number of years before being abandoned for more favorable localities. Although the social organization of early inhabitants is uncertain, by the thirteenth century some Indians were living in large, planned mound communities and were ruled by institutionalized leaders identified as chiefs. Because of European-introduced diseases, coupled with internecine warfare, the Native Americans encountered in and around modern New Orleans by the French colonists at the end of the seventeenth century were a pale reflection of the precontact population.

In the process of utilizing the resources of the marshes, the native occupants transformed them in ways that are both subtle and not so subtle. Perhaps the most obvious means of altering the so-called natural system in the marsh is the existence of archaeological sites themselves.[18] Most sites comprise both deliberate constructions and a less-purposeful garbage disposal. Garbage included the remains of meals, usually fish and animal bone, as well as the discarding of abundant quantities of clams. These clams, known as *Rangia,* comprise an important and common component of archaeological sites in the delta, and their remains mark virtually every site in this region.[19] In many instances, these shellfish remains were so common as to form shell middens or shell heaps. These midden deposits accumulated over different time spans, some long, and some short; some mounds were purposeful and some were not, but all had one common result—vertical buildup of sites above sea level.

The planned and unplanned shell accumulation formed an entirely new ecozone in the marsh. Shell mounds were and, in fact, still are locations of high ecological abundance and diversity. In comparison to the surrounding marsh, archaeological shell deposits manifest a significant difference in the kinds of plants present, as well as their distribution. Since the 1930s botanists have recognized that archaeological sites contributed to the ecological diversity of the marsh and represented unique floristic (and possibly faunal) habitats. These islands in the marsh represent important elements of the modern marsh ecology. After Indians had been virtually exterminated, these shell middens became the locations for historic camps.[20] Even modern fishermen, hunters, and trappers recognize the unique value of the sites as high, dry places to live.

We cannot stop just at the sites themselves, which after all are in part accidental accumulations. The contribution of sites to regional biodiversity and the transformation of the landscape possibly is trivial. The native peoples of this region, however, saw these sites as important elements of the land-scape and reused them through time even after their primary function had ended. The locations of channels and distributary channels of the Missis-sippi River are continually shifting in the delta region in response to a num-ber of natural influences. These shifts are cyclical and are tied to the flow of freshwater in the river and the competition between subsidence and depo-sition.[21] Human occupation of these distributaries was tied to the river's cycle. Occupation of distributaries or branches of the river was remarkably rapid. Precontact groups utilized distributaries as they were forming.[22] These initial occupations were evidently transitory and probably reflect specialized hunting or fishing camps. As a distributary matured, so too did its human use. Mature, stable distributaries were the locus of intensive, long-term sedentary habitations. From these communities native groups hunted and fished, and sometimes farmed, with a concomitant impact on the flora and fauna. Native groups used mass capture techniques to harvest fish, and they intensively exploited local game animals, notably muskrat.[23] While we can-not know for certain what effects these hunting patterns may have had, we do know that the floral and faunal environment were being altered by human actions. As the distributaries began to subside, permanent occupa-tion waned. Because they represented important opportunities for native hunters and collectors to acquire resources that were lacking in the rest of the marsh, these sites continued to be used through time.

A good example of how the shell middens became important habitats in the region can be found in eastern New Orleans, where a series of small natural elevations situated in the midst of brackish water marshes marks the western end of a relict barrier-island trend. These are the so-called Pine Islands. Archaeological sites, principally comprised of midden debris mixed with shell of the clam *Rangia cuneata*, are found on the natural elevations closest to the modern shore of Lake Pontchartrain (see fig. 2.1). Early Wood-land peoples first occupied these sites beginning ca. 500 B.C.[24] These archae-ological sites manifest a very different floral composition today than those elevations that do not support archaeological sites. The Big Oak, Little Oak, and Little Woods sites support stands of hardwood trees, most notably live oak *(Quercus virginiana)*, but also hackberry *(Celtis laevigata)*, cypress *(Tax-odium distichum)*, willow *(Salix nigra)*, and chinaberry *(Melia azedarach)*.[25]

These islands are fringed by marsh elder (*Iva frutescens*), palmetto (*Sabal minor*), and buckbrush (*Baccharis halimifolia*). At Little Oak Island the influence of prehistoric human habitat alteration is most notable. Little Oak Island is one of three "semi-connected sand islands." Its archaeological site is a horseshoe-shaped midden, largely comprised of earth, sand, and midden debris. *Rangia* shells are sparse in comparison to the nearby Big Oak Island site. Immediately south of Little Oak Island is Little Pine Island, with Pine Island to its south.[26] Neither of these latter "islands" was occupied during the prehistoric period. In contrast to the adjacent and physically connected Little Oak Island, the two "Pine Island" elevations support, as their name suggests, a very different flora, especially notable for the presence of stands of pine (*Pinus spp.*) and juniper (*Juniperus spp.*).[27] The presence of nonindigenous chinaberry at Big Oak and Little Oak Islands means that some of the vegetation has been changed in the historic period. However, there seems to be no reason to suggest that the flora of these relict barrier islands is not the result of long-term alteration of the environment, beginning over two millennia ago. These sites ceased to be major occupations after A.D. 200, but archaeological data indicate that they were periodically reoccupied into late prehistoric times.[28] Today, these islands are some of the

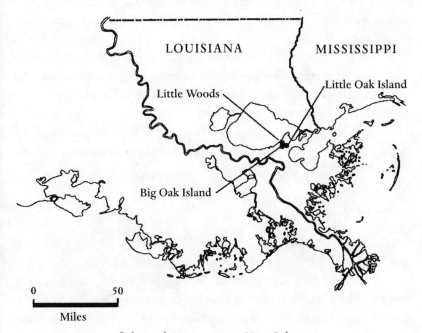

Figure 2.1. Location of elevated sites in eastern New Orleans.

only remaining "natural" habitats left in eastern Orleans Parish. Much of the rest of the region has been transformed for housing developments and industry.

What do these behaviors and impacts have to do with the inevitability of New Orleans as a colonial and later an American city? While some researchers have a relatively low opinion about the resources of the New Orleans area, colonial accounts and maps indicate that many of the most critical resources that made this spot a desirable location are part of the native legacy of the region.[29] For example, the early colonists noted the abundant stands of oak in the region. The Trudeau map of 1803 plots the location of these stands, making special mention of their economic potential.[30] The distribution of these stands of oak parallels exactly the distribution of Native American shell middens. Other examples include the distribution of shell middens along Bayou Barataria, the Lakefront, and Metairie and Gentilly ridges, to name a few.[31] These middens quite literally became part of the infrastructure of New Orleans, as they were excavated to create roads and to line paths.[32] Shells from large middens on Bayou DeFamilles were burned to create lime to plaster the buildings of the emerging city.[33]

The Indians did not physically build the land that would become New Orleans. Their actions over four or more millennia, however, provided resources that made the city's location desirable. Native Americans transformed the marsh by building it up in key locations and by creating islands of biodiversity that were important for hunters, trappers, and fishermen. Native American archaeological sites encouraged the growth of plants and trees that were economically important to European colonists. These same sites were physically incorporated into the infrastructure of the city as shell roads and as lime plaster. Moreover, it was the Indians and their settlements at this particular place that compelled the Europeans settlers to build this "inevitable" city.

BAYOU ST. JOHN

If Native Americans created an actual environment for colonists to use, it can also be said that they further contributed to the "inevitability" of New Orleans by providing environmental and geographic knowledge to the colonists. One of the most crucial pieces of information that the Europeans gleaned from their native counterparts was the location of portages (land crossings between two waterways) from Lake Pontchartrain to the Mississippi River. The most famous of these crossings is Bayou St. John. If there is one element of geographic knowledge that makes this city inevitable, it is this sluggish bayou.

Bayou St. John is a small tidal channel that runs roughly south to north in the mid-city region of modern New Orleans. The geological history of this little channel is relatively simple.[34] Prior to the formation of the modern Mississippi River channel, a river course ran roughly west to east through the center of modern New Orleans. This channel, known as the Metairie Ridge–Bayou Savage course, was abandoned long before the modern city took its geographic form (see fig. 2.2).[35] The relatively high and well-drained Metairie and Gentilly ridges mark this former channel. Geologically, however, the Metairie-Gentilly ridge system forms a barrier between the modern levee of the river and Lake Pontchartrain. Because of the meanders of the modern channel, which impinge on the elevated relict channel, the interior of New Orleans forms a shallow bowl.[36] The modern levee forms its southern, eastern, and western sides, and the Metairie-Gentilly ridge is the northern edge. Periodic flooding of this bowl led to the formation of a drainage outlet, Bayou St. John, which developed at a point where the Metairie-Gentilly ridge was depressed or weakened. Once formed, the bayou served as a natural conduit for drainage from the center of New Orleans. More important for our purposes, Bayou St. John provided an effective portage between Lake Pontchartrain and the Mississippi River.

Archaeological remains at and near the mouth of the bayou demonstrate that Indians knew this portage prior to European colonization.[37] Historical records amplify the archaeological data. In March 1699 the French learned of the portage from the Mississippi to Lake Pontchartrain from an Indian guide. Iberville notes, "The Indian I have with me pointed out the place through which the Indians make their portage to this river from the back of the bay where the ships are anchored. They drag their canoes over a rather good road."[38] This crossing, which utilized Bayou St. John, quickly became the principal route between the Gulf Coast settlements and the new colonies along the river at New Orleans and elsewhere.

There was also at least one other important portage from the Mississippi to Lake Pontchartrain located in modern Jefferson Parish. Sometimes referred to as the Tigonillou (or Tigouillou) portage, or the "Ravine du Sueur," this crossing probably utilized Bayou Labarre and linked up with

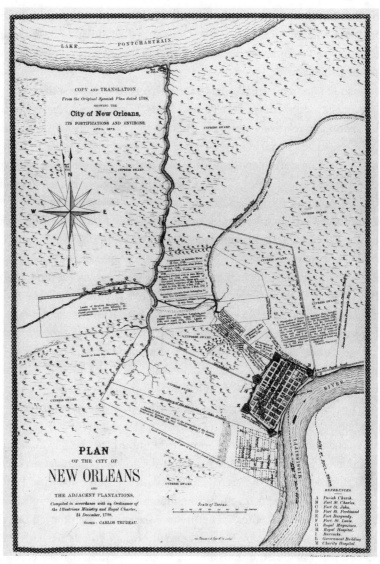

Figure 2.2. An 1875 copy of Carlos Trudeau's "Plan of the City of New Orleans." Bayou St. John runs north to south in the center of the map. Bayou Metarie enters Bayou St. John from the west, while Bayou Gentilly is connected to Bayou St. John via an unnamed channel leading to the east. Bayous Metarie and Gentilly flow in relict channels of the Mississippi River, and their levees form prominent ridges that bisect the New Orleans area from east to west. The extent of the levees can be gauged by the width of the swath of land adjacent to the bayous that is not marked by tree symbols. *(Courtesy of the Louisiana Collection, Howard Tilton Memorial Library, Tulane University.)*

Bayou Metairie near its western end.[39] As late as 1803, Trudeau identified what appears to be Bayou Labarre as the "Tiguayn R."[40] Several large archaeological sites are situated near the mouth of this bayou or are found on or near its banks. Alternately, the Ravine du Sueur could be associated with Bayou La Branche. Here, too, are a number of Native American archaeological sites.[41] A possible third portage was located by Remonville "twelve leagues higher" (upstream) from Bayou St. John. This distance is clearly impossible since it would place the portage well past Lake Pontchartrain, even if the French post league were used as the measurement of distance. Since Remonville also calls this portage the Ravine du Sueur, it is probably the same as the one previously discussed by that name.[42]

Documentary records suggest that these portages were widely used by a number of Indian groups with different tribal affiliations.[43] There is, however, considerable confusion about which Indian group lived near or on Bayou Labarre and Bayou St. John. Iberville, for example, mentions a Quinipissa village near or at the terminus of the portage (probably Bayou St. John), possibly in or near the modern French Quarter. A 1718 map by Guillaume Delisle shows an "Ouma" (probably Houma) settlement at the juncture of Bayou St. John and Bayou Metairie.[44] Villiers, however, suggests that this location for the Houma is in error since they probably lived farther upstream at the time.[45] Le Page du Pratz records that a Cola Pissas (also known as Colapissa or Acolapissa) community once occupied this same location.[46] The Quinipissa and Colapissa may be different names for the same people.[47]

Archaeological evidence for these Indian occupations is, unfortunately, hard to come by. The mining of Indian middens for shell to line roads and paths in eighteenth-century New Orleans and twentieth-century urban sprawl have destroyed much of what was left of the native sites. Recent work at the St. Augustine site in the Tremé district of New Orleans recovered Native American ceramics dating to late prehistoric and early contact times. This site is at or near the actual portage from Bayou St. John to the Mississippi River.[48]

It is possible that these portages were not the locations of permanent habitations in the late seventeenth and early eighteenth centuries. By that time, as a consequence of severe depopulation and internecine warfare, many Native American communities had relocated to sites that offered better defense or more immediate access to European settlements (then located on the Gulf Coast near present-day Biloxi).[49] Native populations in the early eighteenth century evidently were highly mobile and possibly may have had a settlement pattern that included detached parties occupying temporary camps for fishing, hunting, or food collecting.[50]

Bayou St. John is not a historical accident. It is instead an example of the environmental knowledge of the native peoples—knowledge that by its transmission to the French made New Orleans possible. The existence of Bayou St. John is not, however, an instance of natives confronting an overwhelming wilderness. They did not make the bayou or even modify it (to the best of our knowledge). Still, the bayou exemplifies a crucial pattern in the colonization of the New World. Europeans appropriated native knowledge—conscious knowledge—and claimed it for their own. The transformation of this swampy segment of the lower Mississippi River was possible because the Indians had long ago encountered nature and transformed it into a decidedly cultural entity. They remade the land in fact and in the mind, and doing so made New Orleans inevitable.

CONCLUSION

There is little value in dichotomizing humans into noble savages or savage nobles. Such a process does not advance our knowledge of how and why landscapes change or how these landscape changes effect the present. In the context of the New Orleans region, we can suggest two distinct but related aspects of Native American transformation of the environment. First, native peoples did alter the ecology of the region. The transformations may not be quantitatively the same as those in recent times, but qualitatively they are no less real or meaningful. The changed landscapes were both what the Europeans encountered and what shapes our present perception of the modern city and its environs. In addition, Native Americans transformed the landscape of the colonial imagination as well as their actual experiences. The extension of native knowledge to European contexts is what makes Bayou St. John both a tangible image and a metaphor of the historical transformation of New Orleans.

There are no natural contexts in or around New Orleans, and there have not been any since humans entered the region. Accepting all humans—both past and present, Western or native—as agents of change and as part of nature reminds us that the gulf between us and them, past and present, is perhaps not as wide as we think. We can no longer act as if there is a clear distinction between nature and culture. To speak of a natural history requires an understanding of human actions, and to write human history demands an appreciation of nature and its influence. New Orleans is not just a colonial or postcolonial city; New Orleans is the sum of all of its past. We cannot forget that it is humans acting as part of, not apart from, the environment that shape this past.

3

Impenetrable but Easy

The French Transformation of the Lower Mississippi Valley
and the Founding of New Orleans

A S HE PADDLED up the Mississippi River in March 1699, Pierre Le Moyne d'Iberville remarked on the landscape in the vicinity of the future city of New Orleans: "Both banks of the river, almost the entire distance above the sea, are so thickly covered with canes of every size—one inch, two inches, three, four, five, and six in circumference—that one cannot walk through them. It is impenetrable country, which would be easy to clear."[1] The apparent contradiction in Iberville's observation—"impenetrable" but "easy" —expresses both an awareness of the realities of the foreign environment in which he found himself and his unrealistic hopes for the ease with which it might be controlled. For the rest of his days he would struggle to resolve that contradiction, to battle the countryside, to tame it, to possess it, to make it profitable, but, more than that, to make it French. It was a struggle he would not win. Neither would his brother, Jean-Baptiste Le Moyne de Bienville, governor of Louisiana and founder of New Orleans. Neither would any who came after the brothers Le Moyne. Iberville set out an agenda of conquest that began a Sisyphean struggle with nature that continues today.

This chapter will explore some of the interactions between the French and the lower Mississippi valley environment in the eighteenth century. More specifically, it will examine their efforts to create farmland—a European landscape—from cypress swamps and cane breaks. In the first stage of

French settlement, possession of the countryside necessitated working with and adapting to nature far more than they would have liked and in ways not unlike those of the indigenous people whom Europeans displaced. But Iberville's vision of conquest and control persisted, driving the French through a second stage in their relationship with the natural environment of the lower valley, in which they transformed the landscape in an effort to adapt it to them. The founding of New Orleans represented the beginning of a third stage, in which the French sought to create a place that transcended the environment around it. Before a city could even be contemplated, however, the French first had to establish themselves in that impenetrable countryside.

Iberville had it backwards. Penetrating and inhabiting the lower Mississippi environment would prove to be much easier than clearing, draining, leveeing, and transforming it into what he considered a productive land. Living in this environment was easier than controlling it. At first the French would let nature possess them rather than the other way around. In time, however, their willingness—resignation is more accurate—to adapt to a new environment allowed them to plant a foothold in the lower valley, which enabled them to begin to turn the tables, to control nature and break free of its limitations, perhaps, in a sense, to transcend nature.

ADAPTATION

The French arrived in a lower valley already changed by the European presence in America. Chickens walked around the huts of some Indian villages. These domesticated fowl were probably derived from the Spanish, not directly but through Indian trade networks that reached to New Mexico. Fig and peach trees were among the most noticeable European plants that preceded the French into Louisiana, introduced by Indians who traded with the English in Virginia and Carolina. The Spanish had seeded certain places along the Gulf Coast with hogs, which some Indians were incorporating into their diets.[2] The French could not have known that the bison grazing along the north shore of Lake Pontchartrain, along Bayou St. John, and at Tchopitoulas were there in part because the Native American population that had kept the beasts confined to the Great Plains and the prairies of Illinois had declined precipitously since the first Spanish invasions of a century and a half earlier. The French were keenly aware of how Indians had shaped the landscape. At the place where New Orleans would be built, Iberville found trees standing in what he correctly took to be old fields.[3] In 1699 André Pénicault watched some Pascagoulas in the vicinity of Manchac prepare fields for corn and beans, using hooked sticks to pull up cane by its extensive root system, which they left on the ground to dry before setting it afire. He and other French noticed the abandoned fish traps, portages, fields, and villages that marked a landscape altered by people. And farther upriver were the great earthen mounds of the Natchez Indians and their predecessors.[4]

On the one hand, what Iberville described as an "impenetrable" countryside and what Pénicault described as "the excessive growth of trees" suggested the fertility, the richness, the promise of the landscape. Wrote Pénicault, "If the excessive growth of trees with which it is filled were cleared away, the country of Louisiana would be a terrestrial paradise with the agriculture that would be developed there."[5] On the other hand, for that wealth of nature to be possessed, it had to be changed, penetrated, rendered less excessive. There was abundant evidence that change would be easy, or at least possible, because it had happened already. The French sailed into an existing dynamic relationship between people and environment. They would take that relationship in new directions.[6]

The indigenous societies of the lower Mississippi valley had entered a period of tremendous, even traumatic change just before the arrival of the first European explorers, much of it owing to the decline and eventual disappearance of the Mississippian chiefdoms. The powerful and complex societies Hernando de Soto encountered in 1541 were—but for the Natchez —gone by the time René Robert Cavelier, Sieur de la Salle, descended the river nearly 150 years later. Their demise altered patterns of trade, hunting, and warfare from the Ohio River valley to the scattered villages near the mouth of the Mississippi River. Conflict stirred up by European rivalries simply made matters worse, as did the demographic crisis that struck native societies once Old World diseases began to sweep, wave after wave, through valley populations. English allied with Chickasaw, French allied with Choctaw. Many small tribes, termed *petites nations* by the French, were caught in the middle. Seeking to escape conflict with the Chickasaw, a tribe of Tunica relocated from above the Yazoo River to the region of Pointe Coupée only to find themselves very soon at war with the Natchez. Conflict with the Natchez pushed the Tunica into a strong alliance with the French, who arrived on the scene and quickly established their own disastrous relationship with the Natchez. Upon losing perhaps a quarter of their population to smallpox, the Bayagoulas initiated an aggressive policy of incorporating prisoners taken in war and stragglers from other tribes decimated by disease, until a larger, more powerful tribe of Taensas moved into the vicinity. Like the Tunica, the Taensas wished to put some distance between themselves and stronger nations upriver. In so doing they displaced the Bayagoulas as well as a nearby village of Chitimachas. Shifting intertribal relations may also explain the Quinipissas village Iberville found abandoned on a patch of high ground between the Mississippi River and Bayou St. John, the spot Bienville would select for New Orleans.[7]

Initial efforts to create a French landscape failed miserably. The men who would organize the first concessions, the first large-scale plantations along the lower Mississippi, saw as their model the French colonies of Martinique and Saint Domingue, with their large, thriving sugar plantations worked by hundreds of enslaved Africans. They also knew about the English success with tobacco in Virginia and with rice in South Carolina.[8] Or they had in mind the large estates of the French countryside, with their wheatfields worked by a simple peasantry. In a letter to the director of the concession at Pointe Coupée, Eléonore Oglethorpe, Marquise de Mézières —her husband owned the concession—wrote that she hoped he would select a beautiful site for a house, with enough land for three courtyards, a garden, a park, a wooded walkway, and a stream. "One can do as one wishes," she urged, "since the land is ours."[9] One could not simply do as one wished. Louisiana's founders doubted whether sugar could be grown in a place so far north of the tropics. Tobacco seemed more promising, although it could not be grown successfully in the lower reaches of the valley, below Baton Rouge, which is why the French initially put more faith in the Natchez region. Efforts at raising wheat inevitably failed. The climate south of Illinois was too damp for wheat, and it rusted in the fields. Transformation of the countryside into thriving farmland would require not only imagination but also tremendous investments of capital and labor. Without the promise of a valuable staple crop, there was little incentive for such investments.

Nearly ten years after Iberville made his first trip up the river, in 1708 his younger brother, Jean-Baptiste Le Moyne, Sieur de Bienville, made a similar expedition. The passing of a decade had eroded much of the initial optimism for French possession of the lower valley. "This last summer," Bienville wrote the Comte de Pontchartrain, Minister of Marine and Colonies, in

February of 1708, "I examined better than I had yet done all the lands in the vicinity of this river. I did not find any at all that are not flooded in the spring. I do not see how settlers can be placed on this river." Bienville did hold out the prospects of permanent settlement, in a few places perhaps, in particular Lake Pontchartrain and Bayagoulas. "These are the best lands in the world," he reported; nevertheless "I do not see that there is any commerce to be carried on in this country for the present." In 1713 Antoine de la Mothe Cadillac arrived in Louisiana to assume the governorship and noted with much disappointment that reports of the fertility and economic potential of the colony had been greatly exaggerated. Wheat could not be grown at all on the entire continent, he wrote his superiors. Tobacco grew well, only to be destroyed by worms every summer. The pine trees, though plentiful, were useless for ship masts. Corn could not be warehoused longer than a year before it was consumed by vermin and therefore offered no protection against seasons of dearth. Soldiers were dying of scurvy for lack of proper diet, in part because farmers were selling all their vegetables and poultry to the Spanish at Pensacola, who paid in piastres. Cadillac was overly pessimistic. Still, after more than a decade, the French remained unsure about this place, the Mississippi valley, of what they might do with it, and of whether they even wanted it.[10]

And yet, even as Bienville and Cadillac wrote their despairing reports, their countrymen were beginning to settle upon the land. They were doing so, however, not so much by transforming it as by working with it, in essence, by transforming themselves to accommodate the landscape. The first European settlers along the lower river established themselves at the locations of past Indian villages, along Bayou St. John near where New Orleans would be built and upriver at Bayagoulas and Tchopitoulas. Like the Indians before them they planted their crops on natural levees constructed by the river itself. They cleared cane in the same manner as the Indians Pénicault had witnessed. They grazed cattle in the same pastures where buffalo, quickly hunted out and chased away by the French, had grazed just a few years before. Though they continued to attempt to plant wheat after they should have known better, they also planted corn and beans and squash to protect themselves in the certain event that wheat would fail. There were numerous continuities between the French settlers' relationship with the environment of the lower valley and the Indians they

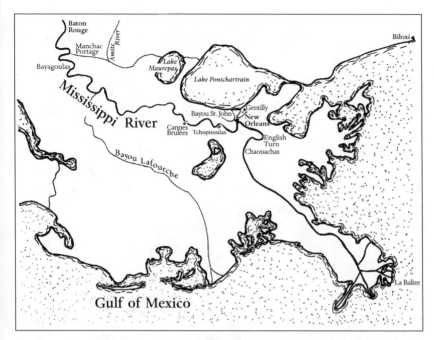

Figure 3.1. French settlements along the lower Mississippi River, c. 1730.

displaced. These continuities permitted the French to win a foothold in the lower valley, which set the stage for later alterations of the countryside.[11]

In 1699 the French first established their administrative base to the east of the Mississippi River, at Biloxi, and then in 1702 at Mobile. By 1715 they were considering relocating to the lower reaches of the river. Mobile was uncomfortably close to Spanish Florida and to English Carolina, although the presence of other European powers would also be reason for maintaining a presence at Mobile. Weighing more heavily in their minds were the problems they were having in providing for their basic subsistence needs. The soil along the Gulf Coast was too thin for sustained agriculture. Meat had become scarce. When settlers first tried their hands at farming the Mobile area "these lands they were extremely productive, which gave them the means to raise poultry and pigs in abundance and enabled them to live rather comfortably; the three last years they have been sterile and have produced neither Indian corn nor vegetables." Moreover, after fifteen years of heavy hunting, wildlife had been much depleted. "The buffaloes and deer are so remote from places where we are established," wrote one official, "that it would cost much more to get it than it would cost to have it brought from France." This situation did not please colonial ministers in France. Removal to the Mississippi River would facilitate trade with Indians and settlers in the Illinois and Missouri river basins, where buffalo and other wildlife remained abundant. The Indians of the lower valley, moreover, never seemed short of corn.[12] Finally, the very places where buffalo had grazed along the banks of the lower river promised to become good pasture for cattle. Still, successful relocation to the Mississippi River depended on the ability of the French to turn a natural environment of lowland prairies, cane breaks, and cypress swamps into pasture and farmland. Dim prospects for this had sent the French to Biloxi and Mobile Bay in the first place. Having found the Gulf Coast unsuitable, they were willing to try settling along the lower river.

In 1708 Bienville wrote his superiors and asked that they arrange to purchase Spanish cattle at Havana, Cuba, and ship them to Louisiana so that herds could be established along the Mississippi River.[13] Commissary General Diron d'Artaguette wrote Pontchartrain that "when there is an abundance of cows meat will be cheap. It requires no trouble to feed them."[14] At that time the Louisiana herd consisted of fifty cows, forty calves, and four bulls, all at Mobile. Between 1706 and 1713, bolstered by regular importations, the herd grew each year at a rapid rate of 30 percent. A similar rate of imports, Bienville and his successors hoped, would provide a base on which to establish a self-sustaining herd. By 1721, as New Orleans was being built, 231 head of cattle ranged along the lower river, mostly at Bayou St. John and Tchopitoulas.[15] A herd this size could sustain an annual slaughter of forty to fifty, but the human population was growing and the herd was too small to serve as an adequate source of meat. Wrote d'Artaguette, "it is desirable that there should be enough beeves to establish slaughter houses in the most populous places otherwise life there will always be rough and very hard."[16] The herd about New Orleans was expected to double by 1725, when it would provide two steers per week for slaughter, enough, d'Artaguette believed, to feed the city.[17]

Of course, the French were not prepared to live on meat alone. The search for a staple crop that would feed a growing European population in Louisiana began with early experiments at planting wheat. Those experiments failed miserably. Corn, a crop native to the Americas and the staple of the indigenous population, became a part of the French diet, of necessity, although they stubbornly resisted its full incorporation into their cuisine. However, when French observers noted seasonal inundations of Indian cornfields, with no damage to that hardy plant, it occurred to them that the region was also eminently suitable to rice cultivation. Again they adapted themselves to the environment by using the flooding power of the river, even as they introduced another new element to it.[18]

Rice first arrived in Louisiana in 1716, brought from St. Domingue, and one or two concession managers sowed fields with it that spring. The first crop must have been very promising, because later that year French officials pushed their most ambitious land-development scheme yet. Following a pattern well established in Quebec, and in anticipation of a land boom, they set up the long lot system along Louisiana river- and lake-fronts. This would keep individual land holdings small and thus discourage speculation and encourage permanent settlement and development. Furthermore, they granted lots on condition that owners bring them under cultivation within two years or face forfeiture. The only agricultural settlement along the lower river in 1716 was at Bayou St. John, where a handful of farmers had been struggling with wheat and living mostly on corn and cattle. They immediately recognized the significance of rice. Earlier settlements had been attempted but abandoned. With the success of the first rice harvest, new settlements quickly appeared at Tchopitoulas, Bayagoulas, Chaouachas, and around Lake Pontchartrain. By 1733 the French were well established in the valley, living and planting behind a rising and spreading manmade wall of earth. Bienville reported, "Three-fourths of the inhabitants live on [rice] and have forgotten wheat bread."[19]

Rice has been a staple of hydraulic agriculture from Asia to Africa and, by 1700, to the coast of South Carolina, and it was very adaptable to the lower Mississippi valley. It withstands heavy rains and flooding much better than corn, itself a hardy plant. While rice was not technically difficult to cultivate, it was rather labor-intensive. A rice farmer would prepare ground with mattocks or sharp hoes, loosening the dirt so it would hold seeds. Once he had sown his fields he would gently flood them until the seeds sprouted and took root. Then he would drain his fields, letting them dry for ten to twelve days so roots and stems could strengthen, while he furiously held back the flocks of birds that eagerly sought to prey on his young seedlings. When shoots reached three or four inches the farmer once again would open his floodgates and let his fields sit under several inches of water, which would shield them from predatory birds and insects and kill strangling weeds. Through the growing season he drained fields as root and stem growth required and flooded them when pests and weeds invaded. In the early fall he would take sickle in hand and harvest his crop. Rice was far more susceptible to weeds than was corn and required constant care through the growing season. Weeding and hoeing in wet fields that bred swarms of mosquitoes was a particularly unpleasant chore. When he completed field work, the rice planter bundled, stacked, threshed, hulled, and stored his harvest, all by hand.[20]

Always holding out the possibility that an environment so abundant in flora and fauna might be possessed and harnessed to their purposes, the French nevertheless found the Louisiana landscape more than a little intimidating. Bienville had chosen not to attack that environment directly but rather to adapt to it, principally by grazing cattle in the very places where he had seen buffalo graze, and by planting rice along the floodplain, as Indians had planted corn. In time his decision proved wise. By 1719 the land from New Orleans to Tchopitoulas and back of the river as far as Lake Pontchartrain was described as one big cattle pasture.[21] Rice agriculture soon spread as well, from English Turn to Baton Rouge. At Cannes Brulées "there are great clear fields which are to be sown" with rice. In one year at Tchopitoulas the concession managed by Sieur de Kolly produced 600 bushels of rice from a mere fourteen bushels sown, in addition to quantities of vegetables and corn. The concessions at Baton Rouge were half burned over and planted with rice and vegetables.[22]

CONTROL

In 1718 the French formally decided to establish a town on the lower Mississippi River. Yet for several years they accomplished little in the way of actual construction as they dragged their feet, waiting to see if such an inhospitable environment could sustain an urban population. In 1720 the seat of government abandoned Mobile but settled in Biloxi rather than New Orleans. That location provided access to the Mississippi River and the trade with Illinois, by way of portages at Manchac and at Bayou St. John. Travel from Biloxi to the river was time-consuming and costly, but none other than a few people stationed at warehouses along the portage routes required regular food shipments.[23] As herds of cattle grew and rice fields spread, French officials gained confidence. In 1720 slaves began clearing land on the natural levee selected as the site for New Orleans. The next year surveyors laid out the town grid. In 1722, only when officials felt certain that the town's inhabitants would shortly be able to live entirely on locally produced beef and rice and thus not depend on costly imports did they move their offices from Biloxi on the coast to the new town on the river. "New Orleans is the place," the Council of Commerce at last declared, that ought to become the center of trade for all of Louisiana.[24]

Surveyors laid out the new town on a natural levee at a sharp bend in the river, in a clearing that had been the site of a Quinipissas Indian village. Like the Quinipissas before them, the French selected the spot because it was the highest and driest for several miles around, and because it was adjacent to Bayou St. John, which offered easy access by canoe to Lake Pontchartrain, the Gulf of Mexico, and the ports of Biloxi and Mobile. New Orleans would not itself become a reliable port for oceangoing vessels until sand bars blocking the river's mouth could be cleared, an undertaking French engineers planned as they built the town. If a town had to be established along the lower reaches of the Mississippi River, and the development of plantation agriculture made that certain, then this was the best location.[25]

Adaptation of agriculture and herding to the lower Mississippi valley environment stimulated more drastic environmental change, exemplified best by the construction of New Orleans. In rural areas, too, adaptation initiated profound and lasting changes. Cattle, for example, though still few in number, destroyed the cane where they grazed. Unlike the buffalo and deer that grazed and browsed on cane shoots periodically as they migrated seasonally over a large territory, domesticated cattle stayed in one place and grazed continuously, leaving the flora upon which they fed no opportunity to recover.[26] The thick mass of cane roots helped hold the natural levees together, permitting the river to deposit soil without washing it away. Where the cane died, the ground captured and held smaller amounts of river sediment, slowing and even halting the natural process by which the river built levees. Land denuded of vegetation was also susceptible to erosion from rain. As natural levees weakened, the river flexed its muscles more often and with great violence. It shifted unpredictably, tearing through and over banks in new places, threatening settlements—originally located where they knew the river's behavior could be predicted—with inundation. This, in turn, encouraged abandonment or demanded leveeing to restore predictability and security. A costly initial investment already made, landowners were not prepared to give up their settlements. In time colonial governments would enforce cattle-grazing restrictions to prevent damage to levees. An apparently simple adaptation to the natural environment—settling where Indians had settled and grazing cattle where buffalo had grazed—led to a chain of events that brought a sharp break with the past and initiated a new relationship between people and the lower valley environment.[27]

The introduction of rice brought similar consequences. Once successfully established as a staple on Louisiana farms, it encouraged farmers to acquire more precise control of the river through leveeing and draining, so that the natural power of the river would be harnessed and used more efficiently. The impulse to construct levees grew more irresistible not only as rice cultivation proved worthwhile but also as cattle weakened natural river embankments. Eventually crops competed with cattle for land and crops won out. By mid-century the inhabitants of the lower valley depended on meat from herds in Illinois and Natchitoches.[28] As with cattle, it was the combination of adaptation and change that permitted the French to penetrate an impenetrable countryside, which in turn led to greater transformations of the sort Iberville had initially envisioned.

In 1723, after a quarter-century of struggle to establish a self-sufficient agricultural economy in the lower valley and Gulf Coast region, colonial administrators declared victory. All but the most recently arrived settlers, they determined, "have sufficient stocks of provisions, so nothing prevents them from producing tobacco and rice" for export. In addition, the Council of Louisiana informed landowners that they would henceforth be required to raise indigo. Landowners resisted these edicts. The efforts of raising a subsistence, they protested, took all their time and energy. Without additional laborers they simply could not be expected to produce a surplus.[29]

The fact of the matter is that settlers would not raise a substantial surplus of rice, whether for sale to nonagriculturists in New Orleans and other towns or for export. To contemporaries they were simply lazy. To historians they possessed a lingering precapitalist mentality that did not respond to the profit incentive. They may simply have weighed the costs of labor invested in planting marketable surpluses against the minimal benefits of trading in an empire notorious for its inability to produce and distribute sufficient supplies of quality consumer goods to overseas colonies. Whatever the reason, settlers would plant food crops, including rice, because they needed them for their own use, although they often had small surpluses for local markets. They hesitated to raise large surpluses of rice for export and could not be enticed to plant staples such as indigo, for which they had no personal use and which only obtained value through very risky overseas trade.[30]

If French settlers could not be made to plant export crops, other people could. In 1719 two slave ships landed at Dauphin Island, near Mobile, the first recorded cargoes of enslaved Africans brought to Louisiana for sale to landowners. Colonial leaders had argued for many years that if Louisiana was to prosper bound laborers would have to be imported. Officials in France always replied that the colony would have to prove its worth before they would permit the diversion of a valuable slave trade from the hugely profitable plantations of Martinique and Saint Domingue. After two decades, that time had come. Cattle and rice not only sustained Louisiana's population but provided the foundation for a promising overseas trade. Between 1719 and 1731 nearly 6,000 enslaved Africans, many of them already experienced rice planters, arrived in Louisiana. Their forced labor enabled the colony to increase rice production at a rapid rate. By 1726 Louisiana was exporting a surplus.[31]

Slaves were not distributed evenly among landowners in Louisiana. They went, of course, to those who could pay for them. Terms of credit were generous, and buyers typically took two years to make their payment. Buyers, however, had to be reasonably sure they could feed their slaves and keep them healthy during the term of mortgage, a period when slaves would strain the current subsistence capacity of the farm. It is not surprising, therefore, that most of the first slave owners were also prosperous cattle herders already producing marketable food surpluses.[32]

The connections between cattle, rice, and slave ownership had consequences for the lower Mississippi valley environment. As slave owners brought more land into cultivation, they sought to protect it from Mississippi River floodwaters. Cattle that grazed on levees and along low-lying grasslands and rice that thrived in periodically flooded fields represented successful adaptations to the environment, so successful, in fact, that they encouraged the importation of slaves. Slaves provided the critical labor for levee construction. Levees made it possible for landowners to transcend the bounds of the natural environment.

TRANSCENDENCE

In 1722 construction on a simple four-foot-high earthen levee began at New Orleans. Others were already well underway at plantations upriver. By 1724 slaves at Tchopitoulas had completed an elaborate system of ditches and levees that stretched nearly ten miles. By constricting the river these levees might well have been responsible for raising the spring flood levels in that year to a new high, flooding New Orleans. City engineers responded by drawing up plans for more substantial dikes made of timber with masonry reinforcements. Four years later the levee at New Orleans remained incomplete, although a section eighteen feet high stood in front of Place d'Armes, now called Jackson Square. In 1731 the city flooded again. Meanwhile, construction along upriver plantation and farm fronts proceeded more rapidly. Several edicts from the Superior Council in New Orleans required all settlers from the mouth of the river to Cannes Brulées, a stretch of about fifty miles, to build levees. Officials in France well understood the source of their profits in Louisiana and prevented their subordinates in New Orleans from drafting slaves from the countryside to work on the city's levee. In 1728 Governor Etienne de Périer wrote the Directors of the Company of the Indies, who managed Louisiana from France, in response to their expressed concerns and assured them that completion of public works at New Orleans would not be allowed to interfere with construction projects underway on the plantations.[33]

By 1732 a levee system was in place, stretching from twelve miles below New Orleans to thirty miles above on both sides of the river. By 1752 the network was extended perhaps another ten miles. If the levees were indeed continuous, and they probably were not until the later date, they could not have been uniform in height and composition. Farmers with few or no slaves would not apply themselves to the arduous task of building dikes. However, farmers who relied only on natural levees, supplemented perhaps with small manmade structures, would have found themselves inundated more frequently by floods that reached record heights each year that their neighbors completed new, more substantial levees, containing greater volumes of water within the river channel. In time such farmers either gave in and bought or hired laborers to build embankments, or else they sold their land to someone who would.

The levee system in place along the lower Mississippi River by the end of the first French period was much more than a wall separating land and water. It was essential technology for productive agriculture. For the fields that stretched out behind them, levees were a mechanism of protection, irrigation, drainage, and fertilization. Unlike modern structures reinforced with concrete, colonial levees were earthen and reinforced by timber. They leaked. During flood stages water seeped through them into a network of drainage ditches that channeled it to low-lying cypress swamps at the back of the lots, where it would collect and eventually drain back into the river downstream. Landowners learned to use seepage to irrigate fields and to replenish them with deposits of fertile silt. Some concession managers stored water within reservoirs so they could flood fields later in the growing season, after the river had receded. By constructing levees, they endeavored not merely to keep the river at bay but to transform it into a tool of agriculture, to use its flooding and fertilizing power to advance their interests as farmers.[34]

The construction of levees changed the geography of individual farms and plantations. Initially landowners cleared, planted, and built their houses and outbuildings on natural levees, the high ground that fronted the river. Over time, as they cleared and drained more of their back lot, they sought to build levees to protect these low-lying areas. To make way for new dikes built on top of the existing natural levee, farmers had to move buildings and other structures back from the river. They also had to clear more land toward the back of their lots. Gradual movement back from the river placed settlers in greater risk of inundation, which in turn necessitated still higher levees and more elaborate networks of drainage canals. Meanwhile, the river rose higher and higher because it was being contained within ever higher and longer artificial walls. In other words, longer, higher, wider levees necessitated still longer, higher, wider levees.[35]

Slavery and levee construction encouraged the French to alter the landscape in another significant way. Agriculture fully occupied slaves only at certain periods of the year, planting and harvest times in particular. During the winter lull, slave owners kept slaves productive by putting them to work on levees and ditches and clearing timber from back lots. Cutting and milling cypress developed into a prosperous off-season industry on Louisiana plantations and provided the colony with another valuable export. By 1729, however, officials expressed concern about the shrinking forests. They suggested that farmers preserve one-third of the timber on their lots, although they granted that this guide would be difficult to follow below Manchac, where wooded land was but a "tongue" between the river and several lakes. Timber for farms in this region, and for New Orleans, would soon have to come from much farther upriver. Of course, deforestation encouraged further expansion of fields into back lots. Once cleared of timber, landowners could not resist the temptation to drain and plant what were swamps, even though by so doing they reduced their capacity to hold levee seepage and floodwater. Moreover, deforestation increased rain runoff because it reduced the land's ability to absorb water, which in turn raised the river level, necessitating more levees and ditches.[36]

Sometimes the French reshaped the environment in ways that defeated their purposes. For example, by draining swamps to prepare for fields, they improved their access to cypress stands, which they could then cut much more easily. However, the easiest way to get timber from the back of lots was to float it out on spring floods. Levees and rice fields meant that timber had to be dragged out by teams of laborers working oxen or else left where it was and burned. Slave owners were in the best position to profit from the market for cypress. Another example is provided by sawmills. The river descended through the lower valley too slowly to make water-powered mills feasible. Timber had to be cut and planed by hand or else by animal-powered mills. Some landowners placed mills near their levees and released high floodwater over the bank and over mill wheels. They did so at their peril. Such practices could weaken levees. The more ingenious planters captured water that ran from the river over the mills, holding it in swamps or reservoirs—provided, of course, they had not turned all their swampland into fields—so it could be used to power mills after floodwaters receded. Late-season milling, however, infuriated landowners downstream, whose fields received much of the sluice. A skillful plantation manager had to manipulate the environment with the precision of a watchmaker: flooding and draining rice fields and timber stands to ensure a timely harvest; allowing sufficient water over the levee to turn mill wheels, but no more than existing swamps could safely hold.[37]

The transformation of cypress swamps, savannahs, natural levees, and cane breaks into cattle pasture, rice fields, and an ever-expanding network of levees and ditches turned Iberville's impenetrable countryside into a fertile and productive colony. Indeed, the environment that had once given the

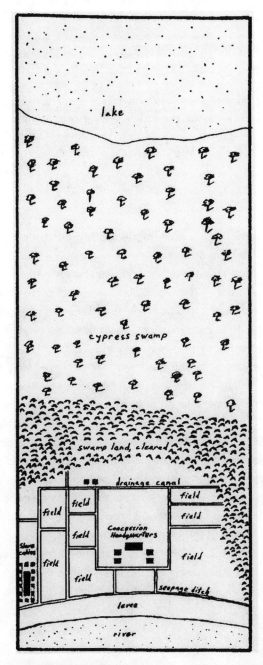

Figure 3.2. Drainage system for a typical French Colonial rice and indigo planta-
tion. After: "Drawing of the Chaouachas Concession," by Jean Francois Ben-
jamin Dumont de Montigny, c. 1730. *Edward F. Ayer Collection, Newberry Library,
Chicago, Illinois, reproduced in Daniel H. Usner,* Indians, Settlers, and Slaves in a Frontier
Exchange Economy: The Lower Mississippi Valley before 1783 *(Chapel Hill: University of
North Carolina Press, 1992), 166–67.*

French pause became the flower of Louisiana. "There will not be any grain this year," wrote Governor Périer in 1730,

> except in the lower part of the colony, since the drought has burnt everything in the upper lands. It was only by irrigating from the river that we have saved the greater part of the rice, which proves that the lower part of the colony will always be more fertile than the upper part, since the fields are easy to drain because of their natural slope; so whether it rains or does not rain, we shall always be assured of the crop when the fields are prepared for bringing the water from the river to them.[38]

The French were now masters of the environment, or so they liked to think. But all they had done, really, was to raise the stakes in their contest with nature. The year after Périer expressed his satisfaction with the landscape the French had constructed, and for several years after that, severe floods broke through levees above New Orleans, submerging more than half the farmland. "If this high water does not withdraw very soon," wrote Bienville in the spring of 1734, "we run the risk of having a shortage of provisions. This would be the third bad year that we should have experienced." He continued, "This country is subject to such great vicissitude that one can almost not count on the crops at all. Now there is too much drought, now too much rain."[39] If the levees and the fields behind them gave the French more control over the environment, they also left them more vulnerable to the vicissitudes of which Bienville spoke. However, a tremendous investment in remaking the landscape had been made; there was no going back. Having initially sought ways to, as Iberville put it, penetrate a hostile environment by adapting to it, the French slowly adapted the environment to them. This transformation proved costly to their colonization efforts and to the natural environment of the lower Mississippi valley. They tried to make nature their servant and instead put themselves at odds with nature, setting up a contentious relationship between people and environment that plagues New Orleans to this day.

CHILD OF THE COUNTRYSIDE

The establishment of New Orleans would represent the most drastic consequence of the alteration of the environment that began with the introduction of cattle and rice. However, New Orleans was the child of changes begun in the countryside. As Bienville wrote in 1723, on moving the government to New Orleans, "It appears to me that a better decision could not have been made in view of the good quality of the soil along the river suitable for producing all sorts of products."[40]

Of course, the town accelerated change in the surrounding countryside by facilitating trade with Europe and the West Indies in slaves, rice, beef, timber, and (later) indigo, and eventually sugar. It also became a market in its own right. Farmers along Bayou St. John took advantage of their proximity to New Orleans and gave up planting grain in favor of truck vegetables, dairy products, and other perishables. Several *petites nations,* including Acolapissas, Chitimachas, and Houmas, moved closer to New Orleans, closer to territory they had inhabited before the French first arrived, and began to supply the townfolk with corn, fish, and game. The moment Bienville decided to move his capital to New Orleans, the chief engineer of the colony contracted with several landowners in the lower valley "to supply planks, boards and beams necessary" to construct new buildings and to erect fortifications at New Orleans and at the Balize. Construction of the Balize, the fort at the mouth of the river, as well as efforts to deepen the channel at that point, only became priorities once the decision had been made to build New Orleans. Throughout the colonial period New Orleans remained dependent on the countryside. The primary current of environmental change flowed from the countryside to the town, although to be sure there was an undertow that flowed back from town to countryside.[41]

For example, the settlements at Bayou St. John and at Tchopitoulas, probably the first places in the lower valley the French identified as suitable for agriculture, grew slowly because the first settlers insisted on planting wheat. The threat of inundation also made landowners skeptical that these places could be settled. The introduction of rice, which made inundation more acceptable, and the growth of cattle herds that grazed along the high grounds on the shoots that appeared as soon as the water receded, were promising and attracted more settlers. By the mid-1720s over 100 people resided at Bayou St. John and over 200 at Tchopitoulas, among them free whites, enslaved Indians, and enslaved blacks, working the land and constructing embankments to control flooding and channel water for rice fields and timber mills. They cut down the cypress. They permitted herds of cattle to consume the cane. They increased the pace of erosion and the likelihood of inundation, which made them even more determined to drain and levee, which required still greater investments in labor, slave labor in particular. At Tchopitoulas slaves built the first extensive network of levees and ditches in the lower valley, a network far more ambitious than the first small earthen dike at New Orleans. Profits in agriculture seemed so promising. Promise encouraged investment in projects to control and harvest nature. Such projects needed protection, direction, and facilitation. That is why the French decided ultimately to establish their capital not at Biloxi, or Mobile, or Natchez, or Baton Rouge, but in the lower reaches of the Mississippi River. For perhaps a century, however, efforts to control nature placed town and countryside in a competition for scarce labor—slave labor—to drain and levee. In this competition New Orleans lost so long as agriculture remained the *raison d'être* for European possession and control of the lower valley.

Throughout the French colonial era New Orleans remained a small, damp, smelly, and often dangerous place to live. Inundation remained a problem through the 1740s, as city engineers struggled to adjust their levee to increasingly higher flood levels wrought by changes in the countryside upriver. A higher, stronger levee kept back the river; it did nothing to keep out water from rain and an occasionally overflowing Bayou St. John. In the low-lying areas toward the back of the town pools of water lay perfectly still but for small dimples made by alighting mosquitoes, imprisoned within an earthen fortress that extended for miles, reflecting clear blue sky above and harboring deadly microscopic organisms within. Into these putrid, stagnant ponds trickled the contents of chamber pots carelessly dumped into streets, manure deposited by the cattle, pigs, and chickens that wandered in and about the town, and the filth from the hundreds of rabbits and doves kept in hutches in nearly every yard. Malaria, dysentery, and yellow fever were facts of life in New Orleans. So, too, were the human remains that ascended to the earth's surface in the town cemetery every time the water table rose. Only the smallness of the population, which did not count 5,000 souls until sixty years after the town's founding, kept diseases from reaching epidemic proportions. Not until the 1790s and even later, in the nineteenth century, as the population grew into the tens of thousands would New Orleans become justly famous for its vicious outbreaks of yellow fever and cholera. Epidemic disease awaited population growth, which in turn awaited the creation and expansion of an agricultural hinterland. That would take time, as it turned out, more time than the French would have. Nevertheless, by 1763, as France handed New Orleans and its surrounding rice and indigo plantations over to Spain, Louisiana's inhabitants had done what Iberville had scarcely dared to dream. They had transformed an "impenetrable country" into a modestly prosperous if not quite thriving landscape of farms and city, an environmental legacy that remains after nearly three centuries.[42]

It is too easy in retrospect to think of New Orleans as inevitable, as predetermined by geography. History suggests otherwise. There was never a major Native American settlement there. Indians harvested the natural environment, they altered it, they participated in trade that reached from the Great Lakes to Mexico. They built towns as populous as important cities in Europe. In all their activity over centuries they built small villages but never a large population or ceremonial center in the lower Mississippi valley. If any place was the inevitable location for a city, it was the region from Natchez to Baton Rouge, with its high ground, fertile soil, and mixed forests. Natchez and Baton Rouge are located at an ecotone, the border of several overlapping ecological zones. Such places typically teem with flora and fauna. They abound in resources that facilitate human settlement. Small wonder it was for so long an important place for Indians.

The French knew about Natchez, of course. And they seriously considered making it their capital.[43] Had they done so they would have followed their own example in Canada, where they built Quebec and Montreal on the sites of Indian towns. Indeed they surely would have made Natchez their capital had it not been for the discovery that rice could be raised in abundance in the lower valley. In other words, had the lower valley not offered the French anything worth investing in, there would have been no need for a port, let alone a capital so far down the river. Natchez or Baton Rouge would have been more suitable. Nothing is inevitable. The success of agricultural activities at Bayou St. John, Tchopitoulas, and elsewhere, which represented the success of the French at winning some control over the environment, in turn created the need for New Orleans.

The changes in the lower valley over the course of the eighteenth century altered the French perspective on the landscape they created. Iberville had looked upon the lower valley with no small amount of trepidation, and with good reason. True, the land had not been impenetrable after all, although possession and control of it had hardly been easy. Still, after sixty years of wrestling with the environment, the French had no regrets. Indeed, it seemed to them that the place desperately needed them. Their alterations of the land left them poised, at long last, to overcome, as one put it, "the natural poverty in Louisiana." The natural poverty. This was not a rich land that the French were harvesting but a poor land that they were improving, the better for all concerned, land and people. That is how they came to see it. This was a confidence that, for better or for worse, would be shared by the Spanish and the Americans each in their turn as would-be masters of the lower valley environment.

Environment
in Service
of the City

The Louisiana Purchase in 1803 greatly enhanced New Orleans's role as entrepôt to the Mississippi valley. Once the full drainage basin became U.S. territory, trade legally moved downstream to the port near the Gulf of Mexico. With the wholehearted adoption of steam navigation during the 1820s, New Orleans's position became even more prominent. Reliable transportation fostered the expansion of the cotton plantation from the lowermost stretches of the river to the Mississippi delta country and upstream to the northern limits of cotton cultivation. With an enlarged cotton-producing hinterland, New Orleans's interests expanded in kind. Where planters sowed cotton, New Orleans had a stake.

The role of the city in shaping the environment became ever more pronounced during the nineteenth century. William Cronon portrays Chicago as a city built on dreams, dreams of becoming the great metropolis of the western United States. To get to that point, Cronon argues, "the land had first to be redefined and reordered." This included removing the Indians, making a wetland into real estate, redirecting the flow of the Chicago River, and funneling commodities through Chicago. Nineteenth-century urban boosters, according to Cronon, understood that cities grew with the countryside, that they "formed a single commercial system."[1] Although Chicago got a later start than New Orleans, it exerted a tremendous influence over its timber- and grain-producing hinterland. Like its northern neighbor, New Orleans, too, was dependent on the growth of its agricultural province and worked to determine more than just the crops planted in Mississippi and Arkansas. It had to manage the transportation corridor that linked it to the rural countryside.

New Orleans and the plantations economically tied to the city shared one cen-

tral concern: the Mississippi River. The river posed problems as well as provided opportunities. In particular, snags could rip the hulls of the wooden steamboats and plagued early transport. To make the river serve the city and agriculturists, steps were necessary to remedy the snag problem. The Army Engineers played a critical role in maintaining the channel by pulling snags. Their task was aggravated by the removal of riparian forests for steamboat fuel and the subsequent erosion of riverbanks, adding more snags to the waterway. Ari Kelman explores this little-discussed dimension of river management and its relationship to New Orleans.

A second critical concern was the seasonal threat of flooding. To contend with the threat of high water, by 1850 individual landowners had constructed levees from New Orleans to Baton Rouge on the east bank.[2] A levee stretched 1,400 miles on the west back up to the Arkansas state line. Planters on the Yazoo delta built levees after the great flood of 1842. The failure of piecemeal levee districts and construction projects became all too apparent during the flood of 1858. Don Davis tells us how frequently flooding occurred and how crevasses, the natural outlet for floodwaters, shaped flood-control practices. New Orleans practically ringed itself with levees to keep high water from inundating the streets. Although early flood-control planners considered using crevasses and artificial crevasses to manage excess water, the dominant protective tool was the levee.

With increasing use of the floodplain for crop production, and severe floods in the 1840s and 1850s, federal involvement became essential to residents of the lower river basin. The coordination of a basin-wide flood protection system hinged on decisions made by federal planners. Thus the debate of the 1840s and 1850s had long-lasting consequences and shaped the massive flood protection system of the 1900s. Ironically, the efforts to protect floodplain agricultural land served to raise the flood levels at New Orleans. George Pabis presents the debate as a backdrop to floodplain concerns in New Orleans.

In their rush to conquer a new land, nineteenth-century Americans sought dominion over nature. Levees along the river were merely one example of a domineering attitude. These chapters represent that sentiment.

Forests and Other River Perils

O N MARCH 24, 1817, a riverboat pilot named Henry Shreve boarded
his steamboat, the *Washington,* at New Orleans's waterfront.[1] Shreve
rushed as he prepared his vessel for a long voyage, fearful his departure
might be blocked by members of the so-called Fulton group, intent as they
were on protecting their exclusive right to power steamboats on the lower
Mississippi.[2] After arriving at the *Washington,* Shreve ordered his crew to fire
the vessel's boilers, and then he steered his craft into the Mississippi's current, bound for Louisville, Kentucky. The pilot kept a log of the trip, noting
that the voyage took only twenty-four days. In less than one month the
steamboat completed the journey from Louisiana to Kentucky, running
against the waters of the Mississippi and Ohio Rivers and into history.[3]

To many observers it seemed that steamboats had finally mastered the
mercurial rivers. In the wake of the journey, Morris Birkbeck, a traveler
making the grand tour of the United States, noted the potential significance
of steam on the major rivers of the Mississippi system. "The upward navigation of these streams is already coming under the controul of steam," he
explained, "an invention which promises to be of incalculable importance
to this new world."[4] For his role in "conquering the rivers," Shreve became
one of the West's great antebellum heroes, a man revered for imposing
order on the mighty Mississippi system, both as an innovator with steam

technology and later as a maverick in clearing dangerous snags from the western rivers.[5]

The voyage of the *Washington* changed not only Shreve's life but, as Birkbeck guessed, would have profound consequences for the whole of the Mississippi valley. Perhaps most important for this forum, in time Shreve's journey proved to be a crucial event in the history of the human transformation of the lower Mississippi. To this point in this volume, the contributors' chapters have demonstrated that the process of shaping and reshaping the environment of the lower Mississippi had been ongoing for centuries when Shreve left New Orleans aboard the *Washington* in 1817. Tristram Kidder and Christopher Morris have illustrated ways in which European settlers and Native Americans alike engineered landscapes for human use in the lower Mississippi valley. Whether in creating edge habitats that fostered hunting, in building levees that harnessed the Mississippi's floodwaters, or in clearing land for agricultural production, human inhabitants of the lower valley attempted to improve their environment, even to control nature. In this light, Shreve's journey stands as one point on a continuum, another step in a long-standing series of human endeavors in the valley.

And yet, the journey of the *Washington* was different from previous transformations of the lower Mississippi. What separated Shreve's accomplishment was its role in launching a new era in which the pace and scale of changes wrought by people in the valley's environment accelerated faster and grew more pervasive than ever before. Shreve's journey aboard the *Washington* helped launch the era of steam on the Mississippi system, a time in which the valley's inhabitants began using industrial technologies in their efforts to mold the world around them.

This chapter will explore the role steamboats played in the human transformation of the lower Mississippi, focusing on three issues: Shreve's role in bringing steamboats to the western rivers, an exploration of the reactions people had when steamboats failed to control the Mississippi system, and an examination of Shreve's actions late in his career. This chapter offers readers a critical perspective on the ways in which steamboats and Shreve transformed the lower Mississippi's environment, moving away from hagiographic portraits typical of the famed riverboat man and romantic viewpoints that often characterize explorations of the steamboats themselves.

THE ADVENT OF STEAM ON THE MISSISSIPPI SYSTEM

From the beginning of the European conquest of the interior of North America visionaries claimed that the Mississippi valley's layout promised riches for settlers in the region, especially those who developed land flanking the lower Mississippi River. The river system, these boosters observed, provided a network of watery paths on which traders throughout the valley would someday float goods to a market located near the entrance to the Gulf of Mexico. Their reasoning was simple and sound: in a period with no paved roads, no rails, no air travel—in short, no technological innovations circumventing the inconveniences of the valley's geography—natural trade routes, and rivers particularly, were the region's commercial highways.

The Mississippi system provided the most magnificent highway system in the world, with over 15,000 miles of tributaries and trunk streams snaking across much of the continent. Early readers of the valley's geography recognized that control of the lower river meant control of the region's trade. One location loomed above others as the key to dominating the river system: the site of New Orleans. After visiting the colony of Louisiana in 1722 and contemplating the Crescent City's natural advantages, Pierre François Xavier Charlevoix wrote, "Rome and Paris had not such considerable beginnings, were not built under such happy auspices, and their founders met not with those advantages on the Seine and the Tiber, which we have found on the Mississippi, in comparison of which, these two rivers are no more than brooks." He went on to predict that "this wild and desert place, at present almost entirely covered over with canes and trees, shall one day, and perhaps that day is not very far off, become the capital of a large and rich colony."[6] Charlevoix was an optimist who based his opinions on a felicitous interpretation of New Orleans's location in the valley.

Predictions such as Charlevoix's hinged on reading the valley's watercourses as uniformly benevolent and ignoring many of the obstacles the region's geography placed on trade and travel—constraints of space and time.[7] For example, Charlevoix's view begged a nagging question: how would traders return upstream against the rivers' currents after bringing their wares to market in New Orleans? In the years prior to the advent of steamboats, the journey from the Crescent City to the upper valley was grueling and dangerous, demanding between three and six months of hard labor in dangerous and foreboding locales.[8] Over land the trail passed through territory that often seemed impassable because of environmental obstacles or the presence of Native Americans hostile to European traders.

The trip upstream via the Mississippi system was usually preferable but could be accomplished only in one of several exhausting ways. In shallow water, riverboat men set long poles in the river's bottom while standing in the bow of their boats and then walked the length of their craft, pushing them along like Venetian gondoliers. In the calmest river eddies oars could be used, and sometimes sails served best in ideal wind and river conditions. Most often, however, cordelling sufficed as the means of defying the river's current. The cordelle was a heavy rope, sometimes nearly a quarter mile long, fastened to the bow of a riverboat. To use it, the best swimmer among the boat's crew swam ashore with the cordelle clamped in his teeth, like a well-trained retriever. Following in his wake came his mates; once on shore they began a months-long game of tug-of-war with the river.[9] Thus, for the whole of the eighteenth century and the beginning of the nineteenth, although the Mississippi system held out great promise for the people of New Orleans, the rivers' currents stood between the city and what many observers perceived as its destiny as the seat of a vast inland commercial empire.

Only in the years following the Louisiana Purchase did boats powered by steam promise a solution to the problem of the current on the Mississippi system.[10] In 1811, convinced that an unproven, relatively new technology could vault New Orleans toward empire, the Legislature of the Territory of Orleans granted one partnership a monopoly of use on the waters of the lower Mississippi River. The recipients, the so-called Mississippi Steamboat Navigation Company, were better known as the Fulton group because Robert Fulton, along with Robert Livingston and Nicholas Roosevelt, formed the company. The legislators offered such a potentially valuable grant in exchange for a promise that the group would bring their steamboat prototype to the lower Mississippi to challenge the river's current.[11] And though the Fulton group arrived later that year, successfully navigating the lower river with their vessel, the *New Orleans,* by granting the partners their monopoly the legislators had unwittingly insured that it was to be years before steamboats were to become common on the river system.[12]

Though the era of steam began when the Fulton group's *New Orleans* made its famous voyage from Louisville to the Crescent City in the winter of 1811–12, the partners' monopoly later dissuaded all but the most steadfast entrepreneurs from experimenting with steamboats on the western rivers.[13]

And the members of the Fulton group, with their monopoly yielding enormous profits, remained content to navigate only the lower—and safest—reaches of the river system, between New Orleans and Natchez, with a tiny fleet of steamboats.[14] As a result, despite the monopolists' successful challenge of the river's current, the majority of the valley's traders did not benefit from the early arrival of steam on the Mississippi system. They still made the long and difficult journey to and from market in New Orleans over land or upriver as they had before the arrival of the steamboats, under their own power. The power of steam remained a local phenomenon, improving the lives of only a few people along the New Orleans–to–Natchez route.

More than five years passed after the first voyage of the *New Orleans* before Shreve shattered the Fulton group's monopoly on the lower river. In the spring of 1817, although the *Washington*'s journey represented few new technological strides, Shreve nonetheless won over lingering skeptics, reassuring them that steamboats would improve the settlers' lives in the valley. One antebellum western historian explained of the *Washington*'s arrival in Louisville that "this was the trip that convinced the despairing public that steamboat navigation would succeed on the Western waters."[15] Shreve's journey aboard the *Washington* was so significant because, by 1817, people had realized that overcoming the rivers' currents represented only a part of controlling the Mississippi system. Though the Fulton group and Shreve himself had proven upstream travel possible, questions still lingered: Could anyone wrest control of the lower river from the monopolists? And, if so, would steamboats come into wide use on the Mississippi system?[16]

Though Shreve's journey aboard the *Washington* did not answer those questions directly, later that spring events following his voyage did. On April 21, 1817, after Shreve's legal counsel squared off with the Fulton group's lawyer in the U.S. District Court in New Orleans, Judge Dominick Hall ruled that his court had no jurisdiction in the case, effectively ending the Fulton group's chokehold on the lower river.[17] As a result, as westerners constructed a regional identity around their region's rivers and the steamboats plying them, they cast Shreve as the hero in the story of steam's arrival on the Mississippi. The role of the Fulton group in the process did not suit regional commentators offended by the monopolists' eastern origins and their desire to close the lower river. Over time, writers created a myth exaggerating Shreve's role in bringing steamboats to the Mississippi system.

They wrote that Shreve alone inaugurated the era of steam, claims that were more an expression of public memory shaped by regional pride than a fair representation of the lengthy process that besting the rivers' currents had been. Shreve had not conquered the Mississippi's currents aboard the *Washington* so much as he had freed the river for public use in the courtroom. For that he deserved his status as a hero and recognition as a key player in beginning the era of steam on the Mississippi.

THE ANNIHILATION OF SPACE AND TIME

When Judge Hall ended the Fulton group's monopoly on the Mississippi system, he started a process in which steamboats were to transform people's relationship with the valley's environment. Nowhere was the impact of Hall's decision more evident than at New Orleans's waterfront, where a technological revolution transpired. The levee at New Orleans presented an awesome spectacle in the years after 1817: a confluence of people, goods, sights, and smells—amazing in an era in which successful technological challenges to the boundaries of space and time had not yet become commonplace.

The key to understanding the transformed waterfront was the gradual arrival over the years of an increasing quantity of steamboats capable of traveling at ever greater speeds. Before 1817 New Orleans's wharf registrar recorded only seven different steamboats arriving at the waterfront.[18] In just over a year following Hall's decision in the Shreve case, fourteen new steamboats appeared on the wharf register, accounting for almost 200 arrivals at the levee.[19] Those numbers marked only the first trickle of what eventually became a flood of steamboats descending on the city. In 1827 there were over 100 steamboats on the western rivers; only five years after that steamboats made over 1,000 separate arrivals at New Orleans.[20] In 1859 more than 250 steamboats traveled the Mississippi system, making more than 3,500 arrivals at New Orleans.[21]

As impressive as the steamboats lined up along the riverfront may have been, many people were more intrigued by the goods those vessels carried to the city and what those goods seemed to symbolize. In the years after Shreve's voyage, New Orleans's levee emerged as an emblem of the fertility of the valley's soil, the industry of its settlers, and the power of technology. Located near the Mississippi system's mouth, New Orleans had always been well situated to serve as a market for the valley, but steamboats brought the promise of the region's geography within the city's grasp much as railroads later linked Chicago and the vast environs of what historian William Cronon has called the "great West."[22] With the arrival of steamboats on the Mississippi system in great numbers, what had been New Orleans's hinterland in theory only became part of the city's zone of trade.

No better evidence of the transformation wrought by the steamboats existed than the towering piles of goods found along the Crescent City's waterfront. In 1816 New Orleans received approximately 4,000 barrels of apples, 38,000 bales of cotton, and 500 hogs, among many other items. The total receipts for the year exceeded $8 million. Impressive numbers indeed, but the following year, after Judge Hall overturned the Fulton group's monopoly, receipts at New Orleans jumped by over $5 million and never again dipped below $10 million. In 1825 the value of receipts climbed over $20 million, and by 1837 that number exceeded $40 million. In that year the mountains of goods found along the waterfront loomed above the surrounding landscape: apples had climbed to 18,856 barrels, bales of cotton to almost 600,000 bales, and pork to almost 9,000,000 pounds. By 1840 over 500,000 tons of freight from throughout the valley had arrived at New Orleans's port, valued at nearly $50 million. Around that time one captivated tourist remarked that "this same levee is the market place of the wealth of the West." Ten years later receipts at New Orleans's port climbed over $100 million and never fell below that mark in the years leading to the Civil War.[23]

Other observers suggested that neither the steamboats nor the produce they carried were the most fascinating sights found at New Orleans's waterfront; instead it was the people who congregated there that best represented the ways in which technology had knit together the vast reaches of the Mississippi valley. Steamboat passengers as they disembarked, dockhands toting loads of cargo, the free people of color and enslaved African Americans working the boats and docks, and a polyglot crowd of traders all gathered along the riverfront.

Varied crowds could be found in other port cities in the United States and throughout the Atlantic world, but the variety of the people at New Orleans's waterfront during the era of steam could be overwhelming. One well-traveled visitor remarked that the "population passing in the streets, especially on 'the Levee,' and others adjoining the river, is the most amusing motley assemblage that can be exhibited in any town on earth." He noted that "the prevailing language seems to be that of Babel—Spanish, Portuguese, French, English, mixed with a few wretched remains of Choctaw, and other Indian tribes; and all these are spoken in the loudest, broadest, and strangest dialects, especially in the markets."[24] Another observer exclaimed that "all grades of society, all classes here mingle & commingle in all the peculiarity of their individual character; as a western buster would say 'stranger, if you want a tall walk & want to see tall sights go for an hour on the levee.' And he who has not seen the New Orleans levee has not seen all of this great country."[25] At the Crescent City's waterfront a multicultural and multinational crowd gathered, brought together in one place by the power of steam.

The scene at New Orleans's waterfront testified to a host of changes steamboats had wrought in people's relationship with their environment. Whiskey from Kentucky distilleries, apples from New York orchards, corn from central Illinois farms, furs from the Canadian backcountry, cotton from the Mississippi delta, starched visitors from London, Creole traders haggling over prices, African American firemen cleaning soot from their faces, so-called Kaintucks napping beside battered flatboats, and genteel couples ambling slowly arm in arm, taking in the sights, all these mingled together at New Orleans's waterfront.

Such a collage of people and goods from lands far removed from one another would have seemed unlikely in the years before 1817, with the lower and upper reaches of the Mississippi valley separated by a half-year's arduous journey. Steamboats made such a gathering possible by transforming New Orleans from a muddy, colonial afterthought in the Spanish and French empires to the center of a vast network of trade, a city seemingly on its way to controlling an empire of its own. Awed by the commercial carnival created by the steamboats he witnessed along New Orleans's levee, one visitor exclaimed, "[N]o triumph of art over the obstacles of nature has ever been so complete."[26] From throughout the Mississippi valley—as far east as the western slope of the Alleghenies, as far west as the eastern slope of the Rockies, and as far north as southern Canada—steamboats brought people and goods to New Orleans, displaying them along the city's waterfront.

How were steamboats able to effect such a transformation in New Orleans's riverfront landscape? The answer lay in the vessels' speed and accessibility as more and faster steamboats plied the Mississippi system with each passing year. After conquering the river's currents, velocity became an all-consuming passion for steamboat men. The boat recording the fastest time on a main route "held the horns" in the parlance of the river.[27] When the *Washington* traveled upstream from New Orleans to Louisville in twenty-five days the feat seemed impossible. Stories recall that when the people of Louisville feted Shreve for his accomplishment, he exclaimed that his record-setting run would quickly drift into the realm of dim memory.[28] Shreve guessed that the journey upriver from New Orleans to Louisville would soon take less than ten days. Witnesses laughed, but Shreve knew his business well. The *Shelby* swept the horns later that fall when it ran upriver to Louisville in just over twenty days. In the spring of 1819 the *Paragon* broke the twenty-day barrier. Shreve's prediction finally came true in 1828 when the *Tecumseh* arrived in Louisville only eight days out of New Orleans.

By 1850 a passenger could leave New Orleans on Sunday for an engagement in Louisville on Friday, confident she would arrive on time.[29] A voyage of nearly 1,500 river miles in less than a week, the distance and time seems paltry and pokey in our era of supersonic travel. For people living in the valley in the years before the Civil War, however, steam travel upended the meanings of time and distance and in doing so altered people's relationship with the Mississippi system and the geography of the valley.

Steam travel collapsed time and space, as a kind of technological alchemy first turned between three and six months' hard labor into one month's comparatively luxurious travel. Less than twenty years later further innovation transformed that month-long voyage into a journey spanning less than a week. So fast were the boats that they appeared able to outpace the seasons. One passenger explained of a late autumn trip down the Ohio and Mississippi that technology had turned back time: "At Pittsburgh the trees are stripped of their leaves by frost. At Cincinnati nature is laying on the last mellow colors of autumn, and the leaves are beginning to fall. At Natchez the forests are still in the verdure of summer."[30]

It seemed as though steamboats had compressed the vast environs of the West like an accordion, bringing the upper Ohio River and the Mississippi delta together with the relative ease with which we might fold a map of the central United States, leaving Baton Rouge astride Pittsburgh. One commentator looking upon New Orleans's waterfront wrote of the boats' impact, "Every day some come from above and others depart, on excursions of one or two thousand miles, to St. Louis, Louisville, or Nashville, or Pittsburgh, or hundreds of other places. For distance is no longer thought of in this region—it is almost annihilated by steam." To the people of the West, and New Orleans particularly, the steamboats represented "a complete revolution in the internal navigation of that region," a revolution that changed the way they perceived their surroundings and their place in the valley.[31]

Observers who walked the waterfront at New Orleans revered the technological innovation that had provided the foundation for the scenes they witnessed. That goods and people from remote lands could meet at the city's waterfront represented a triumph of human ingenuity. As one commentator noted, "The distant points of the Ohio and the Mississippi used to be separated by distances and obstacles of transit more formidable, in passing, than the Atlantic." Then, after the steamboats arrived, "all of the advantages of long rivers remain, divested of all the disadvantages of distance and difficulty of ascent."[32] That the river's current could be overcome, that people could travel in relative ease and comfort upstream from New Orleans to the furthest reaches of the valley seemed to prove that humans could control the river system. One visitor to New Orleans noted that "the former difficulty was in ascending the stream of the Mississippi and Ohio against all the strength and velocity of the current, ever rushing with an overpowering flood to the ocean." In his estimation people, with their invention of steamboats, had cleared those hurdles. "This natural obstacle is now overcome by the inventive genius and mechanical enterprise of man; and the exhalation arising from a few tea-kettles full of boiling water has triumphed over opposing winds and tides."[33]

These revelations tie into a traditional understanding of the role technology plays in shaping people's relationship with their surroundings: in some respects steamboats isolated people from the Mississippi valley's environment, buffering them from the unpredictability of the river system, ultimately diminishing their awe at the power of the nonhuman world in favor of a reverence for new mechanical innovations.[34] As a result, over time, some people expressed greater confidence in humans' ability to control their environment, while others displayed hubris in dealing with the Mississippi system.

NATURALIZING INDUSTRIAL DISASTERS
ON THE MISSISSIPPI SYSTEM

For all of the powerful technology that drove the era of steam, for all of the mechanical, agricultural, and human pageantry on display along New Orleans's waterfront, and for all of the successes people enjoyed in their quest to control the Mississippi system, sometimes steamboats and the rivers they traveled seemed to rebel against their roles as servants of the interests of commerce. Even as people celebrated their victory in stemming the Mississippi's powerful current, the river often displayed its dynamism in ways beyond the power of its waters traveling downstream. River hazards, and snags especially, menaced steamboat pilots and passengers. Snags yanked scores of steamboats into the murky depths of the lower Mississippi, causing huge financial losses and human carnage.

River catastrophes suggested that the Mississippi system could not be tamed as easily as scenes along New Orleans's riverfront suggested, illustrating that even the most powerful technological innovations could not completely protect the people of the lower valley from dangerous elements of their environment. The reactions that survivors and onlookers had to river disasters suggest how people's views of the Mississippi system changed during the era of steam. Residents of the lower valley did not accept that steamboats might sink due to snags found in the rivers. They did not view such disasters as a matter of course in an early period of a new technology's development, nor as limits that their environment had placed on the ease of river transit. Instead, valley residents attempted to consolidate gains they had made in their effort to control the river system by imposing further order on their environment. Their efforts ultimately proved successful in some ways, failed in others, while always yielding some of the most profound human transformations of the lower Mississippi during the era of steam.

By 1823, with nearly 100 boats afloat on the western rivers accounting for more than 750 arrivals in New Orleans, steamboats had become a familiar sight in the lower valley. At that time, the *Tennessee*, a large, handsome craft, was one of the newer boats afloat on the Mississippi. Just after 10 P.M. on February 8, 1823, the *Tennessee*, crammed with passengers and cargo, moved upriver on the Mississippi, just outside Natchez. As a heavy snow fell, the *Tennessee* suddenly shuddered and then came to a dead stop when a fallen tree, with one end anchored in the river's bed and the other rising up to just below the water's surface, pierced the boat's hull.[35] Water gushed into the vessel's innards, and cabin passengers began panicking while deck-hands sprinted around the boat. One account reported that the captain "gave orders instantly to stop the leak; but the pilot, who had been down to ex-amine the damage, with difficulty escaped from the hold, in consequence of the water so rapidly rushing in."[36]

The tree had ripped a hole in the steamboat as large "as a common door," and the "truth was soon told—the *Tennessee* was going down." As the boat sank, some passengers scrambled atop anything they thought might float, while others jumped into the freezing Mississippi, struggling to swim to shore. The crippled boat floated downriver, eventually lodging amid some willows, with only a portion of its topmost deck visible above the muddy water. Reports indicated that at least sixty passengers had died, and for months people throughout the lower valley talked about the sorry fate of the *Tennessee* and its victims.[37]

When the *Tennessee* sank in 1823 the vessel fell victim to the most com-mon of the river perils that plagued the Mississippi system during the era of steam: snags. "Snags" were identified as natural obstacles, seemingly em-bodying a dark side of the Mississippi system. To survivors of steamboat wrecks and observers who discussed them, snags appeared to reach from the bottom of rivers, yanking vessels and passengers into the muddy depths below.

There were two kinds of snags: *planters* and *sawyers,* both products of a dynamic riverine environment. Because the Mississippi created the land sur-rounding it over millennia through deposition, its banks were extremely un-stable and prone to cave-ins. As a result the river's current often tore out whole sections of riverbank, sweeping groups of trees found there, or in some cases even "acres of forests" into the river. Once riverborne, trees ab-sorbed water, losing much of their buoyancy, and some sank to the bottom of the river, bothering no one. Others sank partially, leaving a portion lurk-ing near the surface, threatening river traffic. Trees that became fully fixed in the riverbed were known as "planters" because they became planted in a static position. Others that did not fully plant remained stationary only at their base, while their upper reaches oscillated back and forth in the current, and thus were called "sawyers."[38] Both large planters and sawyers could penetrate the thin hulls of riverboats such as the *Tennessee,* sinking a craft in a manner of minutes.

During the era of steam people learned to fear snags hiding in the river. The valley's residents became terrified and frustrated that even though technology had conquered the Mississippi system's current the western rivers retained elements of unpredictability and danger. Over time, alarmed travelers and river professionals alike elevated snags to mythical status as monsters, likely because the mysterious hazards contrasted to the controlled environment that steam ostensibly guaranteed. Steamboats had promised to regulate travel and commerce, overcoming the Mississippi system's powerful current and controlling its dynamic riparian environment. Catastrophes, in contrast, were unpredictable; they could cause millions of dollars in damage and end hundreds of lives with no warning at all.

Over time people developed what seemed like irrational fears of the river hazards, illustrated in the language used to describe snags. If a snagged tree's canopy remained, some people likened it to a multiheaded hydra. If not, then it could be compared to a long-armed ogre waiting to sink its teeth into vessels. One traveler, the Lady Emmeline Stuart Wortley, contrasted trees on land, with their arboreal splendor, to those in the water, where they appeared as unnatural interlopers: "It is quite curious to see hosts of floating trees, agitated and restless, and ever-tossing about in the rapid current. . . . Who could believe that the birds had ever built and sung in their branches? or that they were appareled in the sweet livery of spring? they have become such black, mummified monsters, and look so hideous and forlorn."[39] So long as trees remained in their "natural" place—on land—Wortley cast a romantic eye on them, but once they found their way into the waters of the Mississippi system they became "unnatural monsters," misplaced threats to commerce and travel.

Both kinds of snags were notorious, but educated travelers and riverboat men reserved special antipathy for sawyers, which often remained hidden beneath the current so "the steersman this instant sees all the surface of the river smooth and tranquil, and the next he is struck with horror at seeing just before him the *sawyer* raising his terrific arms, so near that neither strength nor skill can save him from destruction."[40] Again, here, as in the case of Wortley's observation, the snag, once waterborne, loses its identity as a feature of the "natural" landscape. Instead it is gendered male ("his") and personified further ("terrific arms") as it is cast as a threat.

Sawyers were reviled more than planters for another reason as well. They most often struck boats traveling upstream, because pilots steamed upriver close to shore in the shallows, away from the more powerful mid-stream current, and sawyers frequently reached the surface in the shallows flanking the shore. Additionally, boats traveling upriver were also more vulnerable to the snags because of "the difference of rubbing the back of a hedgehog the right or wrong way."[41] Since disaster most often ensued when boats bucked the river's current—an act some observers believed defied the will of the Mississippi, what they called "nature"—for many people sawyers represented retribution meted out to those unfortunates who dared challenge the river system and by extension the will of nature.

SNAG CLEARING AND DEFORESTATION

Although sometimes couched in irrational language, people's fears of snags were nonetheless founded in bitter experience with the lingering unpredictability of the Mississippi system. A survey done in 1830 found that of the 321 steamboats ever known to have traveled the western rivers more than 10 percent had been destroyed by snags.[42] Of all steamboats falling victim to catastrophe in the years before 1850, snags were responsible for 60 percent of the sinkings.[43]

Faced with the ongoing threat of snags, valley residents began calling for solutions to the problem. The language used in these pleas reveals that people had grown confident in technology's ability to overcome all environmental problems. Drawing on their experiences with the power of steamboats, the valley's residents believed that if the Mississippi's current could be overcome, surely the river's waters could be made safe for commerce and travel. Aware that such an undertaking would be massive and costly, they turned to the federal government for assistance. One memorial to Congress, sent by petitioners from Louisiana, explained: "the difficulties under which . . . the West labors in pursuit of a market for the surplus productions of that fertile region, are twofold—distance from the seaboard, and the danger of the voyage down the Mississippi River. The use of steamboats has annihilated the first; it remains for the Government to . . . remove the latter."[44] As their rhetoric indicated, during the early years of the era of steam the people of Louisiana had become so used to the convenient control of the Mississippi system that steamboats offered that they were unwilling to accept further limitations the river system placed on their commercial pursuits.

Congress replied to such requests by enacting some of the first river-improvement policies in the nation, simultaneously expanding its role in overseeing and shaping the western landscape. As part of an act of 1824 for river and harbor development, Secretary of War John C. Calhoun put out a call—with a $1,000 reward attached—for the best invention designed to remove snags.[45] The bounty went to a Kentucky man whose design fell from use after Henry Shreve offered the government what became known as the snag boat.[46] For years Shreve had been serving as an apostle for new steam technology in the valley. After his journey aboard the *Washington,* Shreve continued trading on the river with innovative, larger steamboats. Aware that while steamboats had bucked the Mississippi's current other environmental obstacles still remained on the river system hindering travel and trade, he designed his snag boat.

The body of Shreve's vessel resembled a steam-powered catamaran; the working guts of the contraption lay in a long steam-powered claw that gripped snags. The claw either plucked snags from the riverbed or raised them high enough so the boat's crew could saw them off, rendering them harmless below the river's surface. To cope with the largest snags, Shreve equipped his boat with a metal beam linking the two hulls at the bow, beneath the waterline. The vessel ran toward a snag, either snapping it at impact, dislodging it from the riverbed, or catching and raising it with the fore beam, at which point the claw plucked it from the river or the vessel's crew sawed it down.[47]

Once they began operating, the snag boats were remarkably successful, justifying the high regard westerners had for Shreve and furthering the impression that human innovation could overcome even the most stubborn environmental problems in the valley. In the span of only one busy month, two of the vessels removed over 6,000 snags.[48] With Shreve's boats working the rivers, financial losses from snags on the Ohio and Mississippi declined from $1,362,500 between 1822 and 1827 to only $381,000 between 1833 and 1838.[49] Commenting on the invention's success, in 1836 a writer at the *New Orleans Bee* claimed that "the Mississippi itself should be improved so far, that in the navigation of it, no obstacles should be encountered except those dependent on weather: for all others—and even those partially—may be removed by art."[50]

As the journalist suggested, many residents of the valley believed "art," or artifice—what we call technology—provided the key to controlling the

Mississippi system, and further technological advance would overcome any remaining environmental constraints on the river system. With the arrival of new technologies *all* obstacles would be removed, including, perhaps, even inclement weather. Such predictions were fanciful, no doubt, but so too had been Shreve's guess that the trip from New Orleans to Louisville could take less than ten days, and that had come true. For many observers the snag boats appeared to have cleared one of the final hurdles on New Orleans's road to commercial empire, as Shreve again offered people a powerful weapon in their efforts at controlling the Mississippi system. If his exploits aboard the *Washington* had played a key role in opening the river system to steamboats, his snag boats went far toward imposing further order on the Ohio and Mississippi, making the rivers safer for navigation.

And yet even as Shreve enjoyed success with his snag boats, the Mississippi system sometimes proved an intractable foe. Snag boats could operate only in seasons of high water; for months each year Shreve watched the vessels languish in dry dock, waiting for rain to raise the rivers' levels. Inactivity frustrated the inventor, who was unaccustomed to giving ground to environmental constraints, and Shreve became overzealous in his response to the planters and sawyers choking the western rivers. He attacked what he perceived as the root of the problem: the trees lining the shore.

In 1827 Shreve stated that the snag problem could only be solved "by cutting down all the timber from off the banks of the river, at all places where they are liable to fall in, from three to four hundred feet from the margin of the river." Later, as he removed whole forests from the riverbanks, Shreve left ample record of the devastation he wrought. In the last quarter of 1832 and the first quarter of 1833 his workers felled 10,000 trees. In his report on the carnage Shreve discounted the protests of his few detractors, stating that "the last named work [felling timber from the banks of the river] is thought by many persons to be an injury, and not an improvement to the river. I am, however, of a very different opinion." Confident in his enterprise, he moved forward. In six months, over the course of late 1833 and early 1834, his production fell to only 1,621 trees removed, but the following year he cut almost 2,500 trees over the same span. Shreve's workers reached full speed between September 1842 and June 1845 when they removed almost 75,000 trees.[51] In Shreve's eyes, not just the snags but the forests lining the Mississippi system's shores had become river hazards.

How had Shreve arrived at such a destructive course? The answer lies in part in ignorance, but also in the ways steam changed people's perception of their environment. As James Hall, a regional commentator, reported, the arrival of steamboats shifted the meaning of forests in the lower valley. Hall wrote, "[T]he immense forests were, before, not only useless, but an obstacle to the rugged farmer, who had to remove them before he could sow and reap. The steamboat, with something like magical influence, has converted them into objects of rapidly increasing value."[52] Hall's observation summarized the manner in which steamboats had transformed the valley's trees from impediments to progress to marketable commodities because of the vessels' insatiable appetite for timber to fire their boilers. Experts estimated that on average steamboats consumed thirty-two cords of wood for each twenty-four hours of running time.[53] More recently, F. Terry Norris has noted that one cord of wood yields approximately 1,500 board feet; thus each steamboat on the Mississippi system consumed more than 50,000 board feet of wood per day.[54] As early as the mid-1820s, when Shreve began cutting timber from the riverbanks, with more than 100 steamboats traveling the Mississippi system the acreage of forest consumed each day for fuel was massive.[55] In short, deforestation was a constant byproduct of the era of steam.

Shreve's actions, however, illustrated yet another way in which steamboats transformed people's perceptions of the valley's forests, as he began cutting trees not for fuel, a productive use, but as a means of keeping them from entering the river and forming snags. In Shreve's eyes the forests lining the rivers' shores menaced trade and transportation on the Mississippi system. He concurred with government inspectors working in the wake of the *Tennessee* disaster who labeled the forests "terrible obstacles" and "causes of much calamity to the people of the West." "Only when the forests shall be entirely cleared," suggested their report, will "these frightful and formidable enemies of western enterprise gradually disappear."[56] Convinced of the nobility of his pursuit, Shreve set about alleviating the problem presented by the dense forests lining the river system's banks.

While Shreve's plan for clearing trees was innovative, it was also fatally flawed and ultimately had devastating consequences for the lower Mississippi's ecology. As he progressed with his work Shreve reassured critics who carped that clearing the banks was a mistake by claiming that "no possible injury can arise from it that I am aware of."[57] Still, despite Shreve's convictions, cutting down the timber lining the Mississippi *was* misguided because the forests had great, though hidden, value in addition to their potential as fuel. Root systems of trees are agents of soil stability. Along the loose banks of the lower Mississippi, trees were often the only thing providing structural integrity to the shoreline. As Shreve removed the timber lining the river, he increased the banks' tendency to cave into the passing waters. Some of his contemporaries also recognized that logging contributed to greater and more frequent flooding in the valley.[58] Floods in turn increased the river's currents, and thereby their corrosive effects on the banks. Again Shreve contributed to the problem he set out to solve.

In light of these facts, it is easy to heap retrospective scorn on Shreve's shoulders. Hubris, we cry; pillager of the landscape; agent of a budding technocracy; and, in some ways, Shreve was guilty on each of these counts. Still, one must remember that Shreve was a product of his times. He was taken with the power of technology to improve people's lives, as were most of his contemporaries who had seen the bright side of the dawning machine age in the West. And, as Shreve explained to critics, there was no shortage of trees in the valley, and he was not alone in viewing a clear-cut landscape as a sign of progress. Perhaps more revealing than Shreve's enthusiasm for deforestation is the minimal resistance found to his program of logging the river's banks. Even as the State of Louisiana enacted legislation to protect riparian proprietors from riverboat men who refused to pay for timber taken during "wooding," there is little evidence of protest against Shreve's actions.[59] This silence likely speaks volumes about the enthusiasm the valley's residents had for steamboats, the man who ostensibly protected them, and federally sponsored river improvements.

CONCLUSION

Shreve's involvement in the advent of steam and snag clearing on the western rivers reveals much about a critical episode in the ongoing human transformation of the lower Mississippi. The era of steam was a linchpin in the valley's environmental history. The arrival of steamboats on the river system marked the juncture of a precapitalist, largely colonial period of resource exploitation, and an era in which people increasingly depended on industrial technologies as they engineered the world around them. Shreve stands out as a crucial figure in that transformation, not just because of his role in opening the river and later making it safer for commerce and travel, but also because of the ways in which contemporary commentators and later historians depicted his role in bringing steamboats to the Mississippi. By reexamining the ways in which steamboats altered people's relationship with the valley's environment, and Shreve's effort to clear the river of snags, we can glimpse how people's relationship with the lower Mississippi shifted in the first half of the nineteenth century. And by debunking romantic myths underlying most work on steamboats, and thinking more critically about heroic portrayals of Shreve, we begin to understand the early impact of technology on the valley's environment.

Deforestation and the so-called annihilation of space and time in the valley are only two ways in which steamboats portended the start of an industrial age on the Mississippi system. With that age came a new era of environmental exploitation and degradation. That people reshaped their environment was nothing new when Shreve set out from New Orleans for Louisville aboard the *Washington* in 1817. During the era of steam, however, people began using more powerful tools as they transformed the lower Mississippi. And as Todd Shallat and Craig Colten demonstrate in later chapters, the valley's residents' growing confidence in and dependence on those tools ultimately led to disasters such as Hurricane Betsy and the rise of a chemical corridor on the lower Mississippi so noxious it is known now as Cancer Alley.

5   *George S. Pabis*

Subduing Nature through Engineering

Caleb G. Forshey and the Levees-only Policy, 1851–1881

FROM THE TIME French engineer Dumont de la Tour outlined the need
for a levee in 1717 to protect New Orleans from Mississippi floods, an
epic struggle has evolved between humanity and nature over control of the
rich alluvial soil along the river. To the founders and early settlers of New
Orleans, levees appeared as the most practical solution to the inundation
problem. For over a century they constructed walls of soil and sand to keep
the waters of the river from intruding onto their property. They were successful, or so it seemed. After 1815 Americans began a campaign of draining
swamps, closing natural outlets, and building more levees along the length
of the Mississippi River. A flood in 1828 spurred another intensive levee-
building campaign. The government of New Orleans and parishes all along
the river passed laws regulating the dimensions of levees, established taxes
to pay for their construction, and prescribed rules for their proper care. By
the 1840s levees extended sporadically from New Orleans to the mouth of
the Ohio River. Despite these efforts, a flood in 1844 broke through to rav-
age plantations in Arkansas, Mississippi, and Louisiana. To rescue civilization from the river, people of the Mississippi delta looked to engineers to
find a solution. How engineers conceptualized the Mississippi River in the
mid-nineteenth century influenced all later engineering projects on the

river. To understand why the Mississippi River today looks the way it does, the story of the engineering debate must be told.

Between 1846 and 1859 several engineers working for the state government in Louisiana expressed to the legislature their concern over the failure of levees to protect people living along the Mississippi River. State engineers Paul Octave Hébert and Absalom D. Wooldridge supported a diversified approach to flood control that depended on the combination of outlets and levees. They criticized current measures that relied solely on levees to prevent overflows. Near-sighted economic interests motivated planters to build levees hastily, blocking outlets and draining swamps without concern for the long-term consequences of their actions. The confinement of the river to a single channel forced sediment that had once accumulated on alluvial lands into the main channel to eventually settle along the riverbed, thereby raising the level of floods. Eventually, inevitably, the Mississippi would overwhelm the levee system and New Orleans would end up under several feet of water. Their critique embodied not merely a call for higher, stronger, and better-maintained levees, it also questioned the present relationship between planters and the river. Floods, they argued, were a natural phenomenon in which the Mississippi expanded its width in order to absorb the greater quantity of water coming down its channel every spring. In Arkansas and Mississippi the river's water seeped into swamps, the natural reservoirs of the river. As the spring floods moved into Louisiana, the Mississippi utilized a system of natural outlets to siphon a portion of its waters into the Gulf of Mexico. By building levees and closing outlets, planters disrupted these natural processes. The result led to ever-higher and more destructive floods. Louisiana's state engineers called for accommodating the river. The only successful method to cultivate the river's alluvial lands involved creating a flood-control system that mimicked nature, and that meant keeping its natural outlets open. By urging the state to adopt new policies aimed at keeping open the last few remaining outlets and building artificial ones to replace those already lost, state engineers tried to temper the drive of many planters to remake the river.[1]

In response to engineers who proposed a diversified system of flood control that included outlets, Caleb Goldsmith Forshey, Albert Stein, William Hewson, and other engineers defended the sole use of levees. Although these engineers may have disagreed on the exact details of what a proper system of flood control entailed, they did share the belief that a diversified program of flood control could only bring disaster to Louisiana. All of them agreed that levees would confine the Mississippi to a single channel and this would force the river to scour out a deeper channel for itself. Levees were not detrimental, but, in fact, vital to any flood-control plan. In 1849 George Willard Reed Bayley, an assistant to the state engineering office of Louisiana, summarized the views of supporters of a levees-only system when he declared, "[O]utlets never will be adopted, they are contrary to the spirit of the age; to that spirit of improvement which would reclaim and cultivate, that would convert every swamp and fen into abodes of wealth, into cultivated fields."[2]

This chapter explores the career of one engineer, Caleb G. Forshey. He might not have been as influential as Andrew A. Humphreys, the head of the Mississippi Delta Survey and later Chief of Engineers of the U.S. Army Corps of Engineers, or James Buchanan Eads, the creator of the Mississippi jetties, in shaping public policy to determine the future of the Mississippi River, but Forshey spent most of his professional life attempting to convince others that only levees, properly designed by engineers, could save New Orleans and the alluvial lands of the Mississippi from flooding. His was one of the most vehement defenses of a levees-only policy in the nineteenth century. Forshey represented the ideas of the majority of engineers who saw levees as the tool to conquer nature and bring a secured prosperity to the people of New Orleans and plantations along the river. This view dominated the engineering community until the devastating flood of 1927. Moreover, Forshey's career reflected the general trends that shaped modern American civil engineering, particularly the increasing professionalization of civil engineers and their conflict with military engineers over the control of the nation's internal improvement projects.

Born in 1813 on a farm in Virginia Caleb G. Forshey spent his childhood in Ohio before entering Kenyon College on a part-time basis. In 1833 his interest in engineering and science spurred him to apply to West Point. Illness forced him to leave the military academy before completing his studies. To regain his health, he moved south to Natchez where he began his life-long interest in the Mississippi River. From 1836 to 1838 Jefferson College in Mississippi employed Forshey as a professor of civil engineering and mathematics. When financial troubles forced the college to close temporarily, Forshey entered the emerging engineering profession.[3] In 1838 the young engineer climbed the high bluffs of Natchez and, as he recalled years later, "looked down upon the Concordia plantations and the vast Mississippi alluvion and deemed it worthy the ambition of a hero in his profession to undertake its rescue."[4] Forshey spent most of the 1840s surveying the Mississippi as an engineer for Concordia Parish. He kept a record of the height of the river near New Orleans which would be very valuable to the Mississippi Delta Survey conducted by the U.S. Corps of Topographical Engineers in the 1850s.[5] In 1841 the Louisiana legislature chose Forshey to be part of a geological survey. He became one of the first members of the New Orleans Academy of Sciences.[6]

In 1850 Forshey's experience and knowledge of the river earned him a key role in drawing up the report of the Senate Standing Committee on Levees and Drainage, the first major study of the Mississippi commissioned by the Louisiana legislature. The committee included in its report several of Forshey's own studies that called for the creation of levee districts and also provided a formula that determined levee dimensions, a survey of all the Mississippi levees in Louisiana, measurements of the Bonnet Carré crevasse, and a survey of the Atchafalaya Basin. Instead of throwing around generalities about the impact levees were having on the Mississippi River, a problem that plagued the engineering debate up to that point, Forshey had measurements to back up his arguments. Unfortunately, his conclusions differed from those of the committee so he had to give his strong dissent in the documents attached to the main report.

In this report and throughout his career, Forshey never wavered in his belief that levees were the only way Americans could reclaim the Mississippi delta. He made it clear that "all levees are closure of outlets; and all outlets, not leveed along their sides, down to the gulf, are but the means of re-submerging the lands which levees reclaim." There could be no compromise with those who wanted to accommodate the river. Outlets represented the opposite of what levees were intended to do, which was to master the river by forcing it into a single channel. His own observations, coupled with other data, revealed that the expansion of the levee system had not raised the Mississippi at all. Levees confined the three to five feet of water that had overflowed 23,000 square miles of lowlands and swamplands safely into a single channel without raising the water level a single inch. If the river had not risen, where did all this water go? To compensate for this greater volume, the velocity of the current increased, which allowed it to scour out a deeper bed. Forshey foresaw no danger in reclaiming the last one-third of the delta still subject to yearly floods. Overflows occurred because of poor levee construction. Clearly, the proprietors living along the river could not themselves bear the burden of building and maintaining a proper system of levees. The devastating floods of the past proved this. Overflows threatened the prosperity of Louisiana. Forshey urged the Louisiana legislature to create a statewide flood-control system, divided into levee districts, an administrative unit whose boundaries followed the natural topography of the Mississippi valley rather than political considerations. An engineer would run each district.[7]

To discover how much improvement the present system of levees needed to meet future floods, Forshey conducted the first survey of the levees ever attempted by the government of Louisiana. He used his own formula to determine how much material engineers needed to make the levees in each district conform to the dimensions he specified. Some parishes, like Tensas, possessed an excellent levee system with a strong tax base to support their maintenance, but most of the other parishes demanded extensive work. Over 847 miles of levees stretched along the Mississippi in Louisiana. Forshey calculated another 159 miles of levee lines would be necessary to fill in the gaps in the present system. To build these levees and upgrade existing ones to the dimensions he recommended, the state needed to spend about $1.3 million. He called on the legislature to have this new system completed in three years. The flood-control system would "thereafter remain a perfect barrier against the attacks of the river."[8]

Breaks in the levees occurred in every major flood. Unfortunately, engineers could only guess what effect these crevasses actually had on the Mississippi. In contrast, Forshey provided the Senate Standing Committee with measurements of the Bonnet Carré crevasse. In the flood of 1849, and again in 1850, the levees just north of New Orleans broke and a rush of water flowed into Lake Pontchartrain, located only a couple of miles away. He measured the length of the crevasse to be 6,804 feet. Nearly three-sevenths of the total volume of the Mississippi flowed through it. Since the crevasse siphoned water from the main channel, Forshey had a chance to assess its effect on the riverbed. His measurements revealed the height of the water above the crevasse to be more than twelve feet greater than directly below. The fall in the volume and velocity of its current created a sand deposit at the downstream end of the crevasse opening. He predicted similar sediment deposits if engineers constructed outlets along the Mississippi. Once engineers sealed the Bonnet Carré crevasse, the return of the normal volume and velocity of the Mississippi current would sweep away the sandbar.[9]

Forshey insisted that only a carefully planned and maintained system of levees could protect the inhabitants of Louisiana. Outlets should be closed. He advised against the clearing of the Atchafalaya because he did not believe engineers could control its waters once they left the Mississippi, as engineers in the twentieth century have realized. In several places along the Atchafalaya the channel lost its velocity, deposited its sediment, and raised the surface of the water.[10] He also recommended closing Bayou Lafourche. Its small capacity made only a small impact on the level of the Mississippi. Keeping this bayou open was not worth the risk of inundating vast areas of rich alluvial lands.[11]

Despite the efforts of the state to solve its flood problems, engineers in Louisiana still looked desperately to the federal government for more information about the river. Luckily, in 1850, Secretary of War Charles Magill Conrad called upon Major Stephen H. Long and the U.S. Corps of Topographical Engineers to survey the Mississippi River in order to explore the best method of protecting the Mississippi valley from flooding. Because of Major Long's responsibilities elsewhere, Captain Andrew A. Humphreys assumed control of the Mississippi Delta Survey. Humphreys was determined to answer the questions posed by the engineering debate raging in Louisiana. He had no predilection for either outlets or levees. Objectivity was his goal. The Mississippi Delta Survey would use scientific methods to measure every aspect of the river pertinent to flood control and provide solutions. Humphreys appointed Caleb G. Forshey to head the hydrometrical party, which studied, among other things, the effect that changes in velocity of the current were having on the amount of sediment carried by the Mississippi. Tragically, during the summer of 1851 Forshey lost his wife to illness but returned to continue the survey soon after.[12]

Pressure mounted on the engineers of the Mississippi Delta Survey to disclose some of their results. The problem of leaking information before a formal report could be given to Congress was a serious one. A controversy arose among the engineers working on the survey when Humphreys found out that Forshey had disclosed data regarding river discharge to George Willard Reed Bayley, who was both an assistant to the state engineering office of Louisiana and an employee of the New Orleans, Opelousas, and

Great Western Railroad, a company with much to lose if the levees could not prevent further damage to its rail system before any official report had been filed with the government. Forshey defended his actions by saying, "[F]ar be it from me to underrate the promise to use for one self the information obtained at the cost, or for the secret advantage of public service. But there is a wide difference between dropping, or uttering an item or two of information, in a time of great public calamity—*for public interest*, . . . and any betrayal of public trust by an employee." He believed the government could not just lock up information that had serious consequences to the public and he felt at liberty to disclose such data.[13] Bayley defended Forshey. He insisted that Forshey only made corrections to a newspaper article. When he asked for more data, Forshey had refused. Furthermore, Bayley explained to Humphreys that the information given to him was minor and that blame should be placed on himself, not Forshey. Unfortunately, as a result of this incident, bad blood developed between Bayley and Humphreys to the point where Bayley warned him, "I have not traversed the valley, and studied the phenomena, of the Mississippi for nothing, during the past fourteen years . . . and I give you friendly warning not to suffer any errors to creep into *your report*. Should you do so, I shall be ever apt to see and criticize them."[14]

Despite these pleas and warnings, Humphreys berated Forshey for his actions. The leader of the survey explained to Bayley that "there is a regular legitimate mode by which those results will be made public, and any partial communication of them by one of the assistants in advance is certain to be detrimental to the rights of others engaged upon that work—to say nothing of the right of government to determine when and how they shall be published if published at all."[15] Humphreys was no less firm in his letters to Forshey, asserting that members of the survey had no rights to the information. It was the government's prerogative to disclose the results of the survey.[16]

Unfortunately, the Mississippi Delta Survey ground to a halt due to the illness of the engineer in charge of the survey, Captain Andrew A. Humphreys, in July of 1851. The suddenness of Humphreys's illness sent confusion among the ranks of the engineers in the field. Forshey represented the feelings of many of his colleagues when he wrote to Major Long "to be informed what are your views of operations, after the date named for the termination of the party's labors, namely March 31st. Are my functions then to cease, and the work indoors and out to be terminated? Or will

I be continued at work upon the accumulated observations, till the means are exhausted, or a new appropriation made?"[17] Forshey sought employment elsewhere. The Mississippi Delta Survey stopped all work in March 1852 and would not resume its operations until October 1857, but without the assistance of Forshey. Andrew A. Humphreys and Henry L. Abbot, who was in charge of field operations after 1857, eventually published their monumental study, *Report on the Physics and Hydraulics of the Mississippi River*, in 1861.[18]

Even if Humphreys did not get along with Forshey personally, he had great respect for Forshey as an engineer. In his *Report*, Humphreys wrote,

> Professor Forshey is entitled to great credit for the zealous and intelligent manner in which he devoted himself, for many years previous to the organization of the Delta Survey, to observing and collecting facts relative to river phenomena, without the aid from any source whatever; he thus accumulated a mass of valuable material, which has been available for the purposes of the Delta Survey. When it is considered how difficult and costly perfect observations are, of the character of some of these made by him as an amateur, it is a matter of surprise that so much should have done by the unassisted enterprise of a private individual. His knowledge of the alluvial region afforded me valuable aid, and I esteemed myself fortunate in securing his services. (4)

However, in his report, Humphreys refuted Caleb G. Forshey's calculations that showed that the development of the levee system allowed the river to deepen its channel. Forshey had gathered the gauge readings of high-water marks at Carrollton and separated them by decade. An average reading for each decade appeared to show that the high-water mark was decreasing as one moved closer to the year 1860. Humphreys noted that the period between 1850 and 1860 saw three years of extremely low water. This had a diminutive effect on the average. If one took the year 1855 out of the average, then the period between 1830 and 1840 had the lowest high-water average. Again, if the numbers were juggled a bit more and the ten-year periods were to begin with 1815 instead of 1810, then the period from 1825 to 1835 exhibited the lowest water while the highest water occurred during the period 1845 and 1855 when the levee system was nearly complete (444).

Before the Mississippi Delta Survey and Humphreys's report, engineers in Louisiana could only speculate on the effect of levees and outlets on the river. Since data had been so incomplete, differing views of nature had a profound influence on the theoretical discussions. If an engineer assumed that every inch of alluvial land needed to be reclaimed, then levees were an obvious choice. On the other hand, if one was sensitive to the relationship between the Mississippi and its swamps and lowlands, then outlets and reservoirs were more favorable. Although Humphreys only saw the river as a series of forces to be studied, he was ready to concede the utility of outlets or reservoirs if data from the Mississippi Delta Survey revealed their beneficial qualities. He questioned their practicability, not on any philosophical grounds but on their cost. Similarly, levees for him were not a symbol of human domination over nature, but merely an expedient solution to a problem. Unlike Forshey, he exhibited no real enthusiasm for them. Because a hard blue clay that the current could not scour covered the bed of the Mississippi River, levees would eventually raise the level of water in the river, Humphreys admitted, but by careful management levees could be built to withstand anything the Mississippi threw at them (24). More than any other engineer studying the problem of flood control, he attempted to be as objective as possible. Unfortunately, the Civil War displaced the nation's resources as its people went from fighting floods to fighting each other. Humphreys's report had little impact on the engineering debate until the 1870s when Humphreys, as Chief of the Army Corps of Engineers, attempted to turn his theories of Mississippi River flood control into federal policy.

In contrast, Forshey's career after the Mississippi Delta Survey hit on hard times. Forshey tried to make a living outside the realm of Mississippi River flood control. From 1851 to 1855 he worked as an engineer for the expanding railroad industry in Louisiana. In March 1855 he moved to Texas, where he opened the Texas Military Institute in Galveston, but the coming of the Civil War forced Forshey to shut it down. Forshey accepted a commission as major with the Confederate forces assigned to defend the Texas coast. Frequent disputes with his superiors led to his dismissal for three months, but the Confederate army reinstated him. Overall Forshey's war record proved quite dismal. Squabbles with his superiors and his belief that his talents were being wasted left him embittered. The war left him without money. When the fighting ended, he moved from one unsuccessful job to another. His luck appeared to change finally in September 1871 when he returned to Louisiana to work for the Louisiana Levee Company and the Board of State Engineers. However, within a few years, the company fired him when he criticized the corruption that ran rampant within it.[19] Moreover, Forshey squabbled with Humphreys over who actually discovered the double-float method used to measure velocity and the idea of collecting matter being pushed along the bottom of the river during the Mississippi Delta Survey. He explained his reason for doing so: "[H]ave I wearied you with my egotism? General, my offspring are few, I am advancing in years, and I would gather them about me, start them off with my mark upon them."[20] Forshey was desperate to leave some legacy despite the failures in his career. But the reality was that Forshey had contributed to the design, while Humphreys had actually suggested the method.[21] When Humphreys heard of this matter, he believed "the claim that he makes is so preposterous that I hardly think he can mean what his words convey, for they prefer a claim utterly unfounded."[22] Humphreys insisted that Forshey only found the most efficient means for fulfilling his instructions. In a letter, after acknowledging Forshey's contributions to the survey, Abbot pointed to evidence refuting his assertions and told him that age had colored his memory.[23]

At the same time as his fortunes appeared to dwindle once more, Forshey received scientific recognition after giving successful papers at the American Association for the Advancement of Science and the American Society of Civil Engineers. Although he never managed to keep a stable job, Forshey continued to write about Mississippi flood control. His ideas about a levees-only policy and his criticism of Andrew A. Humphreys found a welcome reception among the emerging civil engineering community. The post–Civil War engineering debate among engineers on Mississippi flood control cannot be understood outside the context of the emerging professionalization of civil engineering and the alienation of the U.S. Army Corps of Engineers from this process. Self-consciousness among engineers of the importance of their work was clearly displayed in Louisiana during the 1850s. Engineers called upon the Louisiana legislature to give them total control over flood prevention. Only with the expertise of civil engineers could the Mississippi be tamed.[24]

However, members of the U.S. Army Corps of Engineers and the U.S. Corps of Topographical Engineers differentiated themselves from civilian engineers. They modeled themselves on the French: they were educated at military academies, well versed in French and Italian hydraulic theories, and worked as part of a large bureaucratic organization connected with the government.[25] They were part of the emerging professional culture that sought to replace chaos with a system based on scientific data. Military engineers exhibited a mistrust of the English model of civil engineering. In England civil engineers had much in common with the craftsmen of their age: they were independent, trained on the job, and worked on a contract basis. Military engineers believed that their skills and education distinguished them from outsiders who had gained their positions from the reputation of their family or political influence.[26]

Before the Civil War the U.S. Corps of Topographical Engineers had a favorable reputation. A circular in Baton Rouge looked anxiously to the Mississippi Delta Survey, stating that "the reputation of the officers of that corps, will give great weight to the measures they may recommend, and no doubt their report would form the basis of future measures that the Legislature may enact for the protection of the delta of the Mississippi."[27] There were vestiges of dissatisfaction, not so much of the U.S. Topographical Corps of Engineers itself, but of the role of West Point in the education of America's civil engineers. West Point graduated its first civil engineer in 1818. By 1837 over 107 civil engineers had completed their studies at the academy. A steady stream of objections came to the secretary of war questioning the policy of educating more civil engineers than the Army actually needed. Even though most of the engineers working in the field in 1837 had on-the-job-training rather than a formal education at West Point or other engineering schools, some Jacksonian Democrats feared the emergence of a military aristocracy bred at West Point could threaten American democracy. Although this criticism did not lead to any changes, it did reveal a deep suspicion of the government subsidizing the education of civil engineers.[28] Yet such hostility did not translate readily into the area of flood control on the Mississippi. Humphreys and Abbot were both graduates of West Point. The need for the resources and expertise of military engineers in flood control was just too great.

This changed after the Civil War. The U.S. Army Corps of Engineers became the haven of the military engineer, as distinguished from the professional civil engineer. Civil engineers had attempted to organize in 1855, but it was not until 1867 that they successfully launched the American Society of Civil Engineers. The establishment of land-grant colleges in the 1860s ended the near monopoly of West Point. Several smaller institutions educated a whole new generation of engineers who had nothing to do with the military.[29] The number of schools with engineering courses grew from two in

1840 to seventy in 1870. Professional civil engineers distanced themselves increasingly from the military. They respected independence, technical competence, and the social utility of their work.[30]

Forshey's critique of Humphreys and his own recommendations for a levee system must be placed in this context. Unlike Humphreys and Abbott, who spent their entire professional life within the Army Corps, Forshey had moved from one job to another, without any allegiance to any particular institution. In explaining his reasons for leaking information about the Mississippi Delta Survey, he wrote that he had never took "a masonic oath, never to speak of these matters, through all coming time. It would be unreasonable, if not immoral to take such an obligation."[31] Forshey believed firmly that an engineer had an obligation to the public, not a specific government agency, and especially not the military. An examination of three reports by Forshey provides a glimpse of the independent engineer, not only grappling with the problem of overflows on the Mississippi but also confronting the entrenched views of the U.S. Army Corps of Engineers. The issue became who was really more empirical: military or civil engineers.

In January 1873 Forshey and M. Jeff Thompson, as members of the Louisiana Levee Company, submitted their *Report of the Commission of Levee Engineers* to the Louisiana legislature. Two years earlier the state legislature had incorporated the private corporation to begin constructing a flood-control system.[32] The company organized a commission of three engineers to prepare a survey and make recommendations concerning a flood-control system that would be finally able to master the Mississippi. The very need of such a report revealed that the surveys conducted by Humphreys and Abbot after the Civil War had not been extensive enough. The commission gave detailed measurements of all the major breaks in the levee lines on the Mississippi, Red, Atchafalaya, and Quachita Rivers located within the state of Louisiana. Forshey stated explicitly that only by confining the Mississippi to a single channel from the Arkansas River to its mouth at the Gulf of Mexico, closing all the natural outlets, and constructing a levee system according to the measurements recommended by the commission would the alluvial lands bordering the river be reclaimed.[33] However, as earlier engineering reports in Louisiana lamented, the state could not be protected unless Arkansas cooperated in constructing levees that protected northern Louisiana from inundations arising from breaks across the border. Forshey called upon Congress to make such cooperation possible.

The report quotes extensively from a paper Forshey gave at the American Association for the Advancement of Science in Dubuque in August 1872. As with most essays concerning flood control in the nineteenth century, Forshey began with a detailed description of the potential economic benefits that could be derived from the successful cultivation of the alluvial lands of the Mississippi. He concluded, as most Louisiana engineers had for the past two decades, that reclaiming these lands was of the utmost importance for the American economy.[34] Forshey exaggerated that the extremes of heat and cold were unknown in this part of the delta. In making the case of investment in Louisiana, engineers like Forshey never mentioned that New Orleans suffered from yellow fever regularly, with major epidemics two or three times a decade, making it the unhealthiest city in the United States.[35] Forshey concentrated on the positive: close to 20 million acres of productive soil existed for the uses of humanity if only flood control could stem the tides of floods successfully.

At first Forshey acknowledged the contribution of the Mississippi Delta Survey in which he had played a major role. However, he noted, "though most of the questions relating to the physics of the Mississippi proper have been set at rest by the United States Delta Survey, still other problems of not less practical value, whose solutions depend upon long time, are left to be determined by subsequent surveys of like kind."[36] The Mississippi Delta Survey never addressed the issue of how close levees should be placed to the river. The effect of the abrasive waves caused by passing steamboats also needed to be studied. Forshey and several other engineers had made extensive cross-sections of the river. The problem was that the survey, "made by those distinguished laborers, Humphreys and Abbot, has left few benchmarks, and no permanent monuments for reference by the future engineer." No comparisons could be made of the changing nature of the river's banks and its effect on levees. Fortunately, since Forshey had been part of the survey in charge of those measurements, he did remember the locations of several of the more important benchmarks. Forshey called for continual study of the Mississippi. By monitoring the river engineers in Louisiana had made some progress since the Civil War. Water gauges had been established with telegraphic reports published daily. This benefited not only steamboat interests but also gave the people of Louisiana a measurement with which they could anticipate floods by several days. However, Forshey called for more extensive observation of the river. A few engineers recording changes to the river from year to year, he noted, "would do more to solve the occult laws of this giant river than the casual and merely hired services of a hundred machine or routine engineers."[37] A body of engineers assigned permanently to the river could then measure any crevasses in levees soon after they occurred, amassing a log of useful information.

Although the Mississippi Delta Survey cannot be blamed for not predicting the future, it did fail to address the relationship between levees and the emerging railroad network. Forshey was one of the first civil engineers to consider this connection as vital. He explained that "railroads have now become so great a necessity to our civilization that we *must have* them, in addition to navigation, even in the direction of the streams; and when the transverse of the delta of 600 miles long is considered, the necessity is so imperative that all obstacles *must be* removed." Unfortunately for Louisiana, three of the four railroads crossing over the alluvial lands of the state had been devastated by the flood of 1871. Forshey made it clear that "railroads are impossible without levees, and yet civilization and the habitation of the alluvion are nearly impossible without railroads."[38] A properly built system of levees would drain neighboring swamps sufficiently so as to enable railroads to transverse them without danger. He believed that railroad interests, which would benefit greatly from the construction of a properly built and maintained levee system, should support these endeavors financially.

Forshey's report also revealed how vehement levees-only engineers were after the Civil War in condemning all supporters of outlets. He and other advocates of levees had long called for the closing of Bayou Lafourche and the last few remaining natural outlets. However, even as late as 1871 this issue was still in dispute. Forshey explained that

> among the people who live upon the Lafourche and who are interested directly in the matter, it is almost the universal desire to have it closed, and only those oppose it who desire to obtain *cheap coal* for their sugar-houses, or save a few dollars each year on the transportation of their sugar. Possibly some small dealers or trading schooners may oppose it, but fully nine-tenths of the people and property have petitioned for the bayou to be closed.

Besides the economic arguments, Forshey did acknowledge that some planters still believed in the outlet theories for their scientific value. Luckily, he noted, with "reasonable men" he only had to point out that the Lafourche outlet had a cross-section of only 3,500 square feet while the Mississippi it-self had a cross-section of 200,000 square feet. The dangers the bayou posed to planters living along its banks were far more potent than the trifling amount of water that would leave the Mississippi. Even if the Mississippi could be lowered a few inches, other factors called for the closing of Bayou Lafourche: it would be far easier and less costly to maintain a levee 300 feet long that would dike the bayou than maintain the 140 miles of levees neces-sary to keep its waters within its banks. Moreover, the bed of the bayou was rising and would close the channel eventually. To further convince those in the opposition, he quoted Charles Ellet Jr., who had admitted that the Lafourche was the least important of all the outlets on the Mississippi.[39]

The fact that Forshey had to give an opinion on cutoffs provided another indication that Humphreys's report had not solved all the issues relevant to flood control. Forshey and the commission found it necessary to counter the arguments of "cut-off advocates," as he called them. Theoretically, Forshey admitted, cutoffs made perfect sense, but could they be justified economi-cally? By 1872, he noted, twenty cutoffs existed on the Mississippi. Supporters of cutoffs claimed that such cuts in the bends of the river would help lower floods. Unfortunately, no evidence existed that this had happened. In fact, the floods of 1867 and 1871 had reached some of the highest water marks ever known at Vicksburg, Memphis, and New Orleans. Forshey believed that the natural process of making cutoffs represented a temporary shift of a channel. For a time cutoffs may lower the height of floods near its vicinity but at the cost of causing the banks of the river to cave from the increased velocity of the current. The damage to levees far exceeded the benefits from cutoffs. He recommended stopping their use.[40]

Forshey concluded by lamenting that "a want of adequate surveys for our guidance" had inhibited the amount of work the Commission of Levee Engineers had accomplished. The promise of the Mississippi Delta Survey remained unfulfilled, especially for engineers who wanted every outlet closed. Louisiana's engineers had to compile data on their own and wait for further assistance from the federal government.

The need for federal intervention prompted Forshey to write a memo-rial to Congress in 1873 calling for the nationalization of the levees of the Mississippi. In his introduction he explained that he had no economic or political motives for the petition; instead "your memorialist would not rest his prayer upon any grounds but those of reason and science and to these he appeals with confidence that they sustain him in his view."[41] Instead of using materials from the Mississippi Delta Survey, he reprinted much of the information contained in his paper given to the Association for the Advancement of Science. Indirectly, he showed his displeasure with Humphreys. Concerning the Mississippi Delta Survey, Forshey wrote, "[H]aving performed a large share of the labors of the delta survey, to secure which survey my memoir was contributed, with others of analogous kind, and having found my work warmly acknowledged and commended in that most elaborate work, which shall give immortality to its authors" (9). Saying that, he then ignored the recommendations of Humphreys and Abbot and instead reprinted certain parts of his own study written before the Mississippi Delta Survey had gotten underway. It was one thing to submit portions of a paper to Congress that built on the Mississippi Delta Survey, it was quite another to submit a report that Humphreys had dismissed in his *Physics and Hydraulics of the Mississippi River.* Forshey wanted to get the last word in his argument with Humphreys. Originally, Forshey had stated that high-water measurements taken on the Mississippi at the Carrollton gauge just on the outskirts of New Orleans from 1811 to 1850 revealed that the mean high water for the decades beginning with 1811 to 1820 and ending with 1851 to 1860 had not risen, but, in fact, had fallen. He concluded that the river continued to scour out a deeper bed to compensate for the confinement of its waters to a single channel. Humphreys insisted that the decade of 1851–60 had a few years of the lowest water marks ever recorded, which made the mean appear far lower than was warranted. If those low-water marks were taken out, the mean high-water marks were lower for 1851–60 than for the decade preceding it, but higher than the years 1831–40. This revealed that the flood levels were neither rising nor falling.

In his petition Forshey countered this argument by asking, "Why leave out any year? We should take the years as they come. Certain, I would not insist that there might not be a group of consecutive very high waters producing exceptions to this rule, as between 1840 and 1850; not that such exception should not be in the opposite direction" (13). He reminded his readers that this data were not conclusive evidence, but that he had attempted to provide one illustration of what was happening to the river. He stuck to his original observations. He also questioned Humphreys's assertion that the bed of the Mississippi consisted of a hard blue clay. Forshey pointed out that experience in Louisiana had shown that the channels of the Raccourci, Palmyra, and Terrapin cutoffs had undergone changes that showed that the bed of the Mississippi was more fluid than the Army Corps of Engineers had envisioned. He noted that General M. Jeff Thompson, chief engineer of Louisiana, had made diagrams of recent changes to the channel of the Mississippi that showed conclusively how the forces of the current worked easily upon the banks and bed of the river (14). No hard blue clay resisted the current.

Toward the middle of the petition, after making several other critical observations, Forshey added that Humphreys's report "is so thorough and exhaustive as to leave nothing untouched, and the data were spread before the American public and before the scientific world, upon which to base a system of management and control of the floods, and to rescue the alluvium in accordance with the unrestricted ambition of the inhabitants" (14). For Forshey the "war between the outlets advocates and thorough levee men was decided by the Delta Survey in favor of the levee advocates." But then he proceeded to demolish Humphreys's report. Yes, Forshey continued, Humphreys had suggested an outlet near Lake Providence, but he considered the proposal foolhardy. The choice of a location could not have been more impractical. He recalled that during military operations in the Civil War General Grant had attempted to make an outlet at that location and had failed utterly. For him, the last vestiges of the outlet argument had been wiped out and "nothing remains, then, but to put our trust in levees" (14).

However, a successful levee program depended on federal involvement. Forshey made a dramatic appeal to the federal government. The language of the petition was filled with images of impending doom in the struggle of humanity with the natural elements. Before the Civil War Louisiana alone had constructed 740 miles of levees on the Mississippi and 440 miles of levees on the Atchafalaya, Plaquemine, and Lafourche outlets. These levees cost $18 million, with an additional $5 million expended on the levees along outlets. Louisiana, Mississippi, and Arkansas had together expended over $41 million on levees. He noted in the war between the river and humanity, "[I]n addition to this vast sum expended on a conflict of more than 150 years, the loss of more than double this sum has been incurred in the disasters of crevasses and inundations; all wrung from the sweat of a most valiant industrial race, in the case of reclamation and civilization." Such sacrifices demanded government attention. Then, Forshey recalled, "but, alas! The Civil War with its bitter hostilities, came in 1861; and before one year had elapsed, the tramp of armies, with fire, sword, and blood, was upon the people and within two years, these enormous works that had cost such a century of toil were cut down as a military necessity to overwhelm those who resisted the Federal arms." The war brought desolation and, once again, the wilderness sought to take the alluvial lands back from civilization (15).

Attempts to rebuild the levees after the war had exhausted the resources of Louisiana. Forshey estimated that damage from the war required 17,380,120 cubic yards of levees to be rebuilt. People in Louisiana had already constructed 12 million cubic yards. The enormous expenditure had placed a strain on the state's resources. The state could not come up with any more money, so it contracted with the Levee Company to lend it funds. Louisiana's government would pay off the sum in twenty-one years at 10 percent interest. Forshey observed that "in the case of the State of Louisiana the resources for these purposes were wholly wanting. The State Treasury was empty, and the recourse was had to the issue of bonds, for this and many other purposes, proper and improper, till the over-issue of obligations depreciated at the present time, that bonds issued for any purposes are at half their face value" (18).

Forshey saw no other way for the alluvial lands of the Mississippi to be protected than through federal government intervention. The waters moving through Louisiana represented the drainage of twenty-one states and five territories. Over 1.6 million tons of freight in ships moved through the Mississippi in 1871 alone, and every ton of this weight pounded against the levees, damaging them. He asked, "[D]oes it not seem that this was a time for the government to step in and assume the protection of the area rescued from the dominion of the waters. Does not the labor in this conflict, the industrial soldier, whose ancestors for three or four generations have given their lives to this enterprise, deserve repose and laurels for himself and his posterity!" (15). The commerce that traversed the Mississippi from the wealthy states of Missouri, Iowa, Minnesota, Kansas, Nebraska, and Oregon, as well as from several territories, had taken a toll on Louisiana. Forshey argued that it was the responsibility of the federal government to come to the aid of Louisiana. A small tax on the commerce using the river and the railroads of the state would be enough to relieve Louisiana from its burdens (25).

On May 22, 1873, Forshey presented a paper at the fifth annual convention of the American Society of Civil Engineers that was later reprinted in the society's journal. The paper reflected the experience and knowledge of thirty years on the Mississippi. It presented his most concise and articulate statement of the past, present, and future of the levee system on the Mississippi. The paper traced the history of levees, their present condition, and what needed to be done to build a proper system with special attention to the physics of constructing levees. Nothing in the paper was new. The significance of this paper was not what it included, but what it excluded. It made no mention of the largest, most extensive survey of the Mississippi ever attempted, the Mississippi Delta Survey; nor was there a word about the most important report on flood control, Humphreys and Abbot's *Report Upon the Physics and Hydraulics of the Mississippi River*. By contrast, Forshey mentioned the work of William Hewson.[42] The paper exemplified the diminution in the status of Humphreys and the Army Corps in the eyes of civil engineers. The American Society of Civil Engineers looked down upon the military engineers as a lower grade of engineers.[43] In such an atmosphere the fact that Forshey ignored totally the contributions of the Army Corps of Engineers was not a coincidence.

When Forshey died in 1881 after a series of strokes, much had changed in Mississippi flood control. The success of James Buchanan Eads's jetties, despite the pronouncement from the U.S. Army Corps of Engineers that they would not work, added to the public hostility toward Humphreys and military engineers in general. In 1879 Congress formed the Mississippi River Commission, a body of civilian and military engineers, to oversee improvement on the river. Despite Eads's calls to close all the outlets of the Mississippi, the Mississippi River Commission kept the Atchafalaya opened but closed all the others. With this one exception, Forshey's lifelong struggle to have a levees-only policy adopted by the federal government had been realized.

Moreover, by 1881, most civil engineers and military engineers agreed in the idea of severing the natural relationship between the Mississippi River and its alluvial lands. At all costs, they claimed, engineers needed to prevent the river from infiltrating the conquered territory claimed by the American settler. Not surprisingly, they considered the program of the supporters of outlets as dangerous. Americans had rescued the alluvial lands from the river; outlets would only allow the river more opportunities to reclaim them.[44] M. Jeff Thompson viewed those who criticized levees "as visionaries, who do not belong to our age, or period, as geologists might call it, for they propose to levee during the millennium, when 'the lion and the lamb shall lie down together.'"⁴⁵ As Forshey noted, floods "left a desolation over these ten thousand fair fields, and an annual wilderness of water, for several months in the year." Attracting capital became crucial if American civilization was to utilize the many untouched resources of Louisiana. The levees represented the ultimate expression of humankind's domination of the river. Humanity, not nature, would direct the flow of water. Americans had a right to take what they viewed as simply theirs. When engineers established a successful flood-control system Americans could then bring, Forshey predicted, "wealth, refinement, lofty character, and a type of the most exalted civilization known to an agricultural people" to "almost every stream, lake, and bayou, on the delta." All of Louisiana's engineers would have agreed with Caleb Forshey's assessment that "after aiding in all the various and Herculean efforts made by individuals, districts, and States, to accomplish this grand purpose, that the work is too great for either or all these forces; and that unless Government of the United States shall undertake the levee enterprise, the river will, before many years, re-assert his dominion over much of the alluvial region."⁴⁶

It would take several decades and millions of dollars of state and federal funding before levees-only engineers were confident that their work had finally subjugated the Mississippi. But the Mississippi proved Forshey and other supporters of a levees-only policy wrong in 1927 at a terrible cost to the people of the lower Mississippi valley. Forshey must have been rolling in his grave when the U.S. Army Corps of Engineers blew up a levee to create an outlet in order to save New Orleans from flooding in 1927 and later built the Bonnet Carré spillway just above New Orleans to divert flood waters into Lake Pontchartrain and the Caernarvon Freshwater Diversion Structure to flood wetlands in the delta region.

6

Donald W. Davis

Historical Perspective on Crevasses, Levees, and the Mississippi River

> Man-made modifications in Louisiana's wetlands . . . are the
> result of flood protection, deforestation, deepening channels,
> and the cutting of navigation and drainage canals. Reclamation
> and flood control . . . have been more or less a failure, destroy-
> ing valuable resources. Reclamation experts and real estate pro-
> moters have been killing the goose that laid the golden egg . . .
> our future conservation policy should be restoration of those
> natural conditions best suited to an abundant marsh, swamp,
> and aquatic fauna.[1]

A s noted by Percy Viosca in the late 1920s, Louisiana's wetland environments faced a serious dilemma that continues today. The problem
is related directly to humankind's interference with the Mississippi River's
natural flow regime, compounded by the effects of sea-level rise, subsidence,
saltwater intrusion, and erosion created by natural and human-induced
processes. As a result Louisiana's coastal margins and wetlands are vanish-
ing. In these near sea-level habitats, wetland loss exceeds thirty-five square
miles per year,[2] a rate that appears to be decreasing but nevertheless remains significant. When equated with the entire United States, these land
loss assessments are exceptionally high.[3] If they are allowed to proceed un-
abated, in ten years a region larger than the District of Columbia will van-

ish. In less than three decades, Louisiana's land loss will equal an area the size of Rhode Island. In half a century the Gulf of Mexico will repossess coastal property equal to all tracts reclaimed in the Netherlands during the last 800 years.[4]

The culprit in this land to water conversion is human interference with the region's natural hydrology. Louisiana's alluvial wetlands serve as an illustration of how human manipulation of the natural system can drastically affect that system's environmental history. Human elements—specifically those levee-building activities that locked the Mississippi River into a controlled conduit—altered flow regimes, sedimentation, and vegetation. In the process environmental modification contributed to changing Louisiana's coastal landscape. To exist in this dynamic and sometimes inhospitable environment, the population developed and utilized innovative engineering techniques, unconventional wisdom, and unique cultural settlement patterns. The Mississippi River was controlled; land was lost; the environment changed. Now engineers are trying to reconnect the plumbing and mimic the original system. Outlets designed to function like natural crevasses (places where the river overflowed its banks through a narrow, well-defined channel) are being built to divert sediment-laden water into Louisiana's wetlands. We have come full circle.

CREVASSE FLOODING: A NATURAL PHENOMENON

In Louisiana's recorded fluvial history, the interval between 1750 and 1927 can be called the crevasse period (see fig. 6.1). For example, in the fall, winter, and spring of 1858–59 occurred "the worst flood in the history of the Mississippi valley." Numerous crevasses ripped through the levees and devastated the river's bottomlands. More than 25 miles of crevasse breaks submerged this valuable agricultural real estate. H. D. Vogel, in his compilation of Mississippi River data, shows an average crevasse splay encompasses an area of about 650 square miles—the largest involved 2,160 square miles, while the smallest was about 212 square miles.[5] Regardless, a large crevasse can discharge water and sediment over several hundred square miles.

When floodwaters breached a levee, a crevasse furnished a pathway, or vent, for water and sediments to be discharged into the backswamps and basins of rivers flowing from the main channel. This natural process deposited sediment that created a series of deltas on the lower river.[6] When

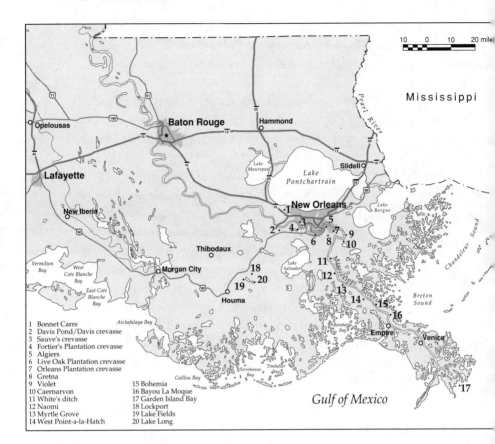

Figure 6.1. Crevasse sites.

Europeans began to manipulate and control the Mississippi River's flow, they subjected themselves, through the crevasse process, to periodic long-term flooding. Once a large crevasse opens, it cannot be repaired until current through the breach subsides as water levels on both sides become equal. When the flow is no longer a raging torrent, usually during the next low-water period, closure procedures are possible. Crevasse flooding was such an acute problem that European settlers took extreme measures to prevent it. Even so, as New Orleans expanded into the surrounding swamps and marshes, flooding ordeals intensified through crevassing. Throughout the early history of New Orleans, few areas within the city were safe from inundation.

NEW ORLEANS: NORTH AMERICA'S
PREMIER BELOW-SEA-LEVEL CITY

The French explorer Jean-Baptiste Le Moyne Sieur de Bienville selected the site for New Orleans despite the objection of Dumont de la Tour, his chief engineer. De la Tour believed the area to be subject to periodic flooding. Indeed, the city could not be built until the water from the 1717 flood receded.[7] De la Tour's beliefs were an omen of things to come, with levees becoming an integral part of the city's geography. In fact, when engineers surveyed Nouvelle-Orleans in 1720 they encircled each block with drainage ditches, establishing the city's dependence on a well-designed levee/drainage network. By 1727 a handmade mud wall protected the earliest city, the Vieux Carre. With time the initial levee extended along the Mississippi River for a total of 42 miles; the first levee built to protect New Orleans was in place. At the same time, the levee was woefully lacking. Since flooding was a recurring problem, the levee was neither high enough nor wide enough. Consequently, high water inundated vast tracts with some regularity (see table 6.1).

The floods of 1735 and 1775 were so high that crevasse waters submerged New Orleans. In 1783 the water "rose to a greater height than was remembered by the oldest inhabitants. . . . [T]he inundation was extreme."[8] From December to June New Orleans remained underwater.[9] In 1785, 1791, 1799, and 1816 floods again submerged New Orleans.[10] In the 1823 flood several crevasses developed between Baton Rouge and New Orleans; one of these was approximately twenty miles upriver from the city and increased long-term flooding within New Orleans.[11]

The river and the city were at war; the river was winning. Even so, Nouvelle-Orleans learned to endure the hardships accompanying the river's annual floods and associated crevasses. The city flooded often, and it responded by making itself a walled compound. Levees constructed by the inhabitants formed protective barriers; without them the Mississippi River would have easily engulfed North America's premier below-sea-level community more frequently. Even so, water often spilled over the levee crests into the city's reclaimed land. When this happened, a second reclamation effort was initiated to pump the floodwater out of the holding basin. Even though the area was drained, the natural system was superseded by an artificial one. At times the artificial system could not accommodate the summers' torrential rainstorms or hurricanes. Therefore, drain and reclaim

Table 6.1. Years for floods of notable importance on the Mississippi River

1543-1799[a]		1801-1927[b]	
1543		1801	1865
1664		1809	1867
1684		1811	1868
1717		1813	1871
1718		1815	1874
1724		1816	1875
1735		1817	1876
1770		1823	1881
1775		1824	1882
1782		1826	1883
1785		1828	1884
1791		1832	1886
1796		1836	1890
1797		1840	1892
1799		1843	1893
		1844	1897
		1847	1903
		1849	1907
		1850	1912
		1851	1913
		1854	1916
		1858	1922
		1859	1927
		1862	

Source: A. G. Warfield, "Mississippi River Levees," in *House Report No. 494, 44th Congress, 1st Session* (Washington, D.C.: U.S. Government Printing Office, 1876); H. D. Vogel, "Annex No. 5, Basic Data Mississippi River," in *Control of Floods in the Alluvial Valley of the Lower Mississippi River, Volume 1, House Document 798, 71st Congress, 3rd Session* (Washington D.C.: U.S. Government Printing Office, 1931); D. O. Elliott, *The Improvement of the Lower Mississippi River for Flood Control and Navigation,* vol. 1 (Vicksburg, Miss.: U.S. Army Corps of Engineers, Waterways Experiment Station, 1932); and R. D. Waddill, *History of the Mississippi River Levees, 1717 to 1944. Memorandum* (Vicksburg: Mississippi River Commission, 1945).

[a]Discounting 1543, 1664, and 1684 the average is one flood every 5.8 years.

[b]From 1801 to 1927 there was a flood, on average, every 2.6 years.

measures are partly responsible for the levee-protected city being nearly eighteen feet below sea level. At least 45 percent of the metropolitan core is at or below sea level.

Levees and drains protect the populace and infrastructure. A single pump failure, levee crevasse, hurricane, or a twenty-four-hour rainfall event of ten to fifteen inches can be disastrous. Levees are essential to keep flood-water out; they keep rainwater in as well. This water is, in most cases, discharged by pumps that function continually to remove surface runoff and groundwater seepage. Since much of the drainage network is gradually being updated, any breakdown necessitates immediate repairs. If pumps fail during a major storm, the city's topographic "bowl" floods, compounding the disaster.

When it was established in the early 1900s the drainage network was adequate and considered excellent. Although more than 200 years old New Orleans was still in its infancy; major population increases had not developed (see table 6.2). After 1920 the alluvial wetlands, much of it near or slightly below sea level, began to be reclaimed. The problem was how to drain them in a cost-effective manner. Gravity flow was impractical; thus a heavy-duty pump was the solution. Designed by New Orleans engineer A. Baldwin Wood, the pump quickly revolutionized New Orleans's urban geography. Areas that were inaccessible were opened to settlement, but swamp drainage was a major undertaking. Pumps, canals, and levees had to be built or installed. Soils "were not soils at all, but a thin gruel of water and organic matter that shrank and settled when the water was removed."[12] Twenty-two drainage-pumping stations drain New Orleans. Several are considered to be the world's largest, with the pumping capacity to empty a ten-square-mile lake, 13.5 feet deep, every twenty-four hours.

Prior to 1955 developers opened only a few subdivisions in the former bald cypress (*Taxodium distichum*) and tupelo gum (*Nyssa aquatica*) swamps. They had built most older residential areas on natural levee deposits or re-claimed Lake Pontchartrain marshes. Housing contractors built homes with raised-floor foundations, supported by masonry blocks that could be raised and releveled. If this technique could not be utilized, wooden pilings were driven under the slab to a depth of at least forty feet. Pilings were necessary. Since the soils are poorly suited for urban uses, these construction methods withstand subsidence well.

Table 6.2. Population of New Orleans, 1820-1990

Year	Population
1820	27,176
1830	46,082
1840	102,193
1850	116,375
1860	168,675
1870	191,418
1880	216,090
1890	242,039
1900	287,104
1910	339,075
1920	387,219
1930	458,762
1940	494,537
1950	570,445
1960	627,525
1970	593,471
1980	557,927
1990	496,938

Source: Louisiana Almanac, 1995-1996 (Gretna, La.: Pelican Publishing, 1995).

CREVASSES: A REPRESENTATIVE SAMPLE

Captain A. A. Humphreys and Lieutenant H. L. Abbot, topographical engineers, described the flood of 1849 as the most destructive flood known. The first great crevasse occurred in March a few miles south of Red River Landing on the Mississippi's west bank, while others occurred throughout the drainage basin. These breaks remained open and submerged much of the Atchafalaya basin to the west. At Brashear (now Morgan) City water was over the banks for eight days. Fifteen miles north of New Orleans, at Fortier's plantation, a large crevasse flooded the country between the Mississippi River and Bayou Lafourche to a depth of four feet. On the east bank Sauve's crevasse (eighteen miles above New Orleans) did an "immense amount of damage," flooding the city for forty-eight days.[13]

In 1850 a crevasse more than one mile wide broke through the levee north of New Orleans at Bonnet Carré (also in 1849, 1871, and 1874) (see table 6.3). During each of these floods water flowed through the break for

Table 6.3. Bonnet Carré Crevasse summary

Date of crevasse	Maximum width	Closed
December 1849/1850	6,700 ft.	1850
April 1859	450	1859
April 1871	2,400	1872
April 1874	1,370	1883

Source: C. Ellet Jr., "Report on the Overflows of the Delta of the Mississippi," in *Senate Executive Document, No. 26, 22nd Congress, 1st Session* (Washington, D.C.: U.S. Government Printing Office, 1852); *Memorandum Relative to Bonnet Carré* (Vicksburg: Mississippi River Commission, 1931); and G. Gunter, *The Relationship of the Bonnet Carré Spillway to Oyster Beds in Mississippi Sound and the Louisiana Marsh, with a Report on the 1950 Opening and Study of Beds in the Vicinity of the Bohemia Spillway and Baptiste Collette Gap* (New Orleans: U.S. Army Corps of Engineers, 1950).

more than six months at an estimated rate of 150,000 cubic feet per second.[14] The 1874 break remained open until 1883. For ten years water flowed through the gap, raising the level of Lake Pontchartrain by two feet.[15] In addition, the floodwaters swept away "dwellings, sugar-houses, crops, and fences, like chaff before the wind, tearing out railroad embankments for many miles, interrupting railroad communication from the north with New Orleans, and even threatening the safety of the city itself."[16] Since the opening functioned for ten years, hundreds of fishermen were unable to harvest Lake Pontchartrain's oyster beds and saltwater fishery. Freshwater saturated the lake's brackish water, dramatically altering the habitat. Navigation was difficult, and the lake trade, which employed at least 3,000 vessels and supported nearly 40,000 individuals, faced potential ruin.[17]

A crevasse about a mile north of Lockport in 1851 remained open for months and reduced flow in lower Bayou Lafourche by about three feet.[18] Five additional crevasses were reported south of Lockport.[19] Local citizens believed this problem was directly related to obstructions placed in Bayou Lafourche by General Andrew Jackson during the War of 1812. Levees at this time did not extend south of Lockport.[20] Consequently, an 1854 crevasse flooded the backlands into Lake Fields and Lake Long west of Bayou Lafourche. Contemporary observers estimated water would flow through this levee break for at least five months.[21] In 1858 two crevasses were reported south of Lockport on Bayou Lafourche. The Bell and La Branche crevasses broke through the three-foot-high levees and crested seven feet above the yearly mean.[22]

Table 6.4. Crevasses on the Mississippi's west bank between the Red River and Head of Passes, 1882 to 1912

Year	Number of crevasses
1882	20
1884	12
1890	22
1892	7
1893	4
1897	1
1903	4
1907	1
1912	3

Source: *Annual Report for the Mississippi River Commission, 1915* (Washington, D.C.: U.S. Government Printing Office, 1915).

The flood of 1865 produced fifty-nine crevasses, one of which was two miles wide and submerged thousands of acres.[23] From 1866 to 1874 more than 100 miles of Louisiana's levees collapsed into the main channel.[24] The most destructive flood to be recorded on the Mississippi River up to that time occurred in 1882. Crevasses associated with this flood "left the people of the valley prostrate." At New Orleans flood stage lasted ninety-one days. In 1882 284 crevasses were reported on the Mississippi River (see table 6.4); in 1883 the number was 224.[25]

After these floods many citizens contemplated moving out of the flood zone. They were afraid the levees would be abandoned and not repaired. The Mississippi River Commission began closing crevasses associated with previous flooding events, appeasing citizens' fears, and thus began the levees-only policy. According to now General Humphreys and civil engineer Charles Ellet this was not the best policy. Nevertheless, the levees rose. The Mississippi River Commission did not consider outlets or reservoirs; they built only levees. An engineered channel incarcerated the river and forced sediments through the modern delta into the Gulf. This process also systematically deprived Louisiana's marshes and swamps of the sediments considered vital to keep the environment near equilibrium.

The U.S. Army Corps of Engineers eventually adopted the 1890 flood as the "standard project flood." It served as a benchmark for new levee-

Table 6.5. Louisiana crevasses in 1912

Location	Date	Length	Cause of break
Salem	April 12	4,975 ft.	Sand boil
Lower Tensas	April 12	No data	No data
Torras[a]	April 12	3,800	Structural weakness
Hymelia[b]	May 1	4,000	Sand boil
Bayou Sara	May 3	190	No data
Cannon	May 14	60	No data
Within the Atchafalaya Basin			
Moreauville	May 6	10,000	No data
Alto	May 19	1,200	No data
Bayou McCracken	April 6	5,300	No data
Bayou Atkins	April 10	5,300	No data

Source: A. Stickney, O. H. Ernst, H. M. Adams, T. L. Casey, H. L. Marindin, R. S. Taylor, and J. A. Ockerson, *Annual Report of the Mississippi River Commission* (Washington, D.C.: U.S. Government Printing Office, 1903); and "Levees," in *Report of the Mississippi River Commission, Appendix MMM* (Washington, D.C.: U.S. Government Printing Office, 1912).

[a]Flooded 1,345 square miles.

[b]Flooded 1,310 square miles; also flooded in 1903.

design criteria.[26] Nevertheless, thirty-eight crevasses broke open in the 1897 flood. More important, the Mississippi River Commission proclaimed that "flood waters can be permanently controlled by a system of levees."[27] Levee construction and maintenance proceeded as the most common flood-management tools; consequently, the 1903 flood resulted in only seven crevasses. The flood of 1912, the largest in recorded history at that time, spawned twelve crevasses between Cairo, Illinois, and New Orleans.[28] Ten of these were in Louisiana (see table 6.5).

Levees constructed by the federal government sustained little damage during the flood of 1922. The only problems were two breaks south of New Orleans. At Myrtle Grove a crevasse 1,000 feet wide flooded the west bank, while a break at Poydras (now Caernarvon) contributed to east-bank flooding. The break, 1,500 feet wide, resulted in a wall of water 115 feet high—as tall as an eleven-story building—exploding through this crevasse.[29] It was clear the levee system was beginning to control the Mississippi's floodwater, at least until the flood of 1927 changed the nation's opinion on flooding.

THE EVOLUTION OF THE MISSISSIPPI RIVER'S LEVEE SYSTEM

In flooding, velocities decrease away from the main channel. Slower-moving waters deposit heavy and coarse sediments on the river's bank. Reccurring floods have elevated the Mississippi's banks, producing an easily defined natural levee system.[30] Flanking the Mississippi River, these natural levee ridges, or topographic "highs," are ideally suited for agriculture and settlement. The height and width of these features depend on the size of the river, stream, or bayou that created them. The larger the water body, the wider and more pronounced the natural levee it created. The region's first immigrants recognized that the elevated, fertile natural levees were capable of being quite productive. Overbank flow was a serious problem, so to protect their investment planters began to build crude, flimsy mud levees. Unfortunately, these earthen ridges were so small and low that floods frequently overtopped and broke them with ease.[31]

After the 1828 flood, which produced the highest water recorded up to that time, the population requested better levees.[32] By 1851 most of the levees between New Orleans and the Red River had risen to at least four and a half feet—the largest was eight feet, with a base of thirty-three feet.[33] With time these embankments extended from New Orleans to the mouth of the Arkansas River, a distance of about 600 miles and prevented flood-waters from flowing into the backswamp reservoirs.[34]

By current standards all levees constructed in this period were insufficient in height and width. They were inefficient and failed because of overtopping, crevassing, seepage, and bank collapse. It was a weak system, easily breached, allowing crevasse-induced flooding. An indication of the levee's deficiency is the number of crevasses identified with each flood (see table 6.6). Many of these breaks were the result of poor maintenance, poor planning, and the plantation community's lack of vigilance and experience.

Newspaper accounts note many crevasses were a product of defective rice flumes placed within the levee (flumes siphoned water off the main stream to help flood the rice fields). Holes were punched in the levee to guarantee a near-constant water supply. These siphons were weak points in the levee and caused considerable problems. State and parish (county) officials were responsible for insuring these rice flumes were safe from accidental rupture. Despite an allegedly regular inspection system, rice flumes

Table 6.6. Crevasses on the Mississippi, principally in Louisiana

Flood	Crevasses
1773	1
1775	1
1782	1
1785	1
1791	1
1799	1
1816	1[a]
1823	1[a]
1847	1
1849	2
1850	8
1851	8
1858	45
1859	32
1869	16
1874	43
1882	284
1883	224
1884	204
1890	53
1892	31
1893	17
1897	37
1903	6
1912	15
1913	8
1916	1
1920	2
1922	4
1927	13

Total number of crevasses reported in the literature is 1062.

Source: A. B. McDaniel, "Flood Waters of the Mississippi River," in *Senate Document No. 127, 71st Congress, 2nd Session* (Washington, D.C.: U.S. Government Printing Office, 1930); H. D. Vogel, Annex No. 5, "Basic Data Mississippi River," in *Control of Floods in the Alluvial Valley of the Lower Mississippi River, Volume 1, House Document 798, 71st Congress, 3rd Session* (Washington D.C.: U.S. Government Printing Office, 1931); D. O. Elliott, *The Improvement of the Lower Mississippi River for Flood Control and Navigation,* vol. 1 (Vicksburg: U.S. Army Corps of Engineers, Waterways Experiment Station, 1932a); and R. D. Waddill, *History of the Mississippi River Levees, 1717 to 1944, Memorandum* (Vicksburg: Mississippi River Commission, 1945).

[a]The literature mentions several crevasses, but does provide a single figure.

repeatedly served as break points for crevasses. For example, a derelict rice flume allowed a crevasse at Live Oak Plantation in Plaquemines Parish in 1882. The plantation's levees were "absolutely rotten, [collapsed] in places for more than a mile and letting in a flood which overflowed . . . the towns of Algiers and Gretna."[35] Obviously the inspections were not thorough enough.

Two years later Davis crevasse (on the west bank, approximately twenty miles north of New Orleans) was directly attributed to an old rice flume.[36] The crevasse was

> by far the most destructive single crevasse known in the history of Louisiana. Its waters extend in a sea of more than a hundred miles along the west bank of the Mississippi river. St. Charles, St. John and St. James parishes . . . are wholly or partially submerged . . . portions of Jefferson, Orleans and Plaquemines parishes . . . are under water. Thirty or 40 miles away . . . the great flood spreads out like a modern deluge to the banks of the Lafourche, above Thibodeaux.[37]

From this experience scientists concluded "rice flumes are the great crevasse makers, and should be abolished."[38] Farmers did not heed the warning of 1884, and rice flumes continued to initiate crevasses through the late 1800s and early 1900s.[39]

THE SWAMP LAND ACTS OF 1849 AND 1850

> For hundreds of miles I found the whole bank of the stream under water; [the water] was rushing over the country with the force of a torrent carrying ruin and destruction in its course. A great number of plantations and settlements were entirely submerged.[40]

From 1840 to 1860 a series of harsh floods made it apparent the flooding problem was too large to be handled at the local level. It was a national issue. Much of the unprotected/unoccupied land was in the public domain. If reclaimed and protected against overflow, the land would become a national economic asset.[41] Two severe floods in 1849 and 1850 convinced the federal government to provide financial assistance to construct a continuous levee system. Additionally, an 1850 survey reported Louisiana's privately built levees were about five feet high and three feet wide at the top sloping down to a 7.2-foot base. To expand this network was beyond Louisiana's capabilities. The state needed major financing.

The federal government recognized the financial problem and enacted the Swamp Land Act of 1849.[42] This act applied only to Louisiana, while the Swamp Land Act of 1850 extended to other states.[43] Through these acts, Congress granted each applicable state the swamps and overflowed lands within their boundaries. Congress dedicated proceeds from the sale of this land to levee construction and reclamation projects.[44] The goal was to make these lands fit for cultivation and settlement. Louisiana, Mississippi, Alabama, and Missouri organized offices to sell their newly acquired swamplands. However, lack of coordination and cooperation among states prevented the immediate construction of an effective levee system. The Swamp Land Acts sought to build a contiguous levee but failed in this respect. Even in the strictest sense, the acts were not a flood-control measure. After some time, however, they did mark the beginning of federal participation in flood-related issues.

While the federal government deliberated the merits of flood control, the states created levee districts and levee boards. These quasi-public corporations received the authority to construct the drains necessary to reclaim the swamps and overflowed lands and to fight the battle against floods. With these boards and the federal government working together, by 1858 more than 2,000 miles of levees up to ten feet high, with a base up to seventy feet, outlined the Mississippi River. Louisiana had 745 miles of levees—a system that cost the state $18,000,000.[45]

The Civil War left this lengthy levee system in shambles due to intentional destruction or neglect.[46] Crevasse after crevasse further demolished the postwar levees. By 1878 "hundreds of miles of the main line had disappeared or had been destroyed."[47] The problem was so severe that some landowners seriously considered abandoning the levees and allowing the river to overflow its banks and flood the formerly protected fields. Riparian landowners were losing to the river. Even so, the federal government showed greater interest in the river. The detailed and scholarly A. A. Humphreys and H. L. Abbot Delta Survey was published in 1861. In 1865 Secretary of War Edwin M. Stanton, commissioned General Humphreys to investigate the status of the Mississippi River's levees. After Humphreys determined their status, Congress wanted to know what it would cost to repair the levees. Even with this federal interest, Reconstruction delayed effective rehabilitation. Levees needed to be repaired, but funds were unavailable. For a time, the river regained control, and natural processes dominated. Though interest in reclamation and flood protection waned, it did not disappear completely.

Floods in 1868 and 1874 rekindled the levee issue. Congress reconsidered support for flood-control legislation, particularly when it was reported after the 1874 flood that "gaps in the levees equaled from one-third to one-half of the entire length of the levees."[48] After this flood Congress convened a board of engineers—the Levee Commission—to assess the flooding problem. Five years later Congress created the Mississippi River Commission, which immediately put the United States into the flood-control business. In 1882, based on "the theory that confinement of flood water would periodically flush out the channel, thus removing obstructing bars and preventing the formation of new obstructions,"[49] the commission adopted a levee policy.

Creation of the commission marked the end of the nonsystematic approach to levee construction. The era of uncontrolled, unstructured, and unsupervised construction was over. The levees were defective because they had been built without effective supervision or coordination from a central authority.[50] The new principle called for consistent design and construction standards that would lessen the defects and produce a more efficient levee system.

The Mississippi River Commission altered levee construction. Early levee specifications called for careful clearing of the right-of-way. Builders closed root holes, excavated parallel drain ditches, and used the spoil to build the embankment. "Station men" using wheelbarrows worked 100-foot sections of levee. As the levee increased in height and volume, mules and scrapers shaped the bulwarks.[51] With time, coal, gasoline, diesel, fuel oil, or electric dredging equipment replaced mules.[52] The machine-made levees often were set back and raised in order to maximize high-water protection. By 1901 most of the levees along the Mississippi as far as New Orleans averaged twelve feet in height. In some stretches they were over twenty feet high. South of New Orleans the levee was between five and six feet high.[53]

By 1927, from Cape Girardeau to New Orleans twenty-eight levee boards controlled riparian lands.[54] Thirteen levee districts gained incorporation in Louisiana and had the power to levy taxes, issue bonds, and generally do whatever was necessary to construct and maintain levees within each district. In addition to the money generated internally from each district, the proceeds of a one-mill tax throughout Louisiana was available for levee work.[55]

RECLAMATION AND THE SWAMP LAND ACTS

Reclamation was an important part of the Swamp Land legislation. Local newspapers presented south Louisiana's swamps and marshes as "nearly valueless until reclaimed by drainage."[56] In the 1800s "it was believed by properly managing rainwater runoff and high water the whole of this coun- try [south Louisiana] may be reclaimed and made in the highest degree pro- ductive."[57] The levee issue was so acute the state engineer in 1854 reported that along Bayou Lafourche the "amount of swamp and overflowed land, [which is] perfectly worthless, would, after being saved from inundation by [levees], sell for more than enough to pay for [them]."[58] Furthermore, recla- mation of the overflowed lands would eliminate the habitats preferred by mosquitoes carrying yellow fever. This was a latter-day argument used to promote reclamation efforts.

Between 1880 and 1930 reclamation projects intensified within Louisi- ana's alluvial wetlands, most notably in New Orleans.[59] As a result, flood- and water-control methods extended the cultivated or settled land beyond the natural levees down the slope into the backswamps. Riparian land- owner holdings expanded considerably as a result. During this period, more than fifty land reclamation efforts transformed over 13,800 acres.[60] Recla- mation entrepreneurs felt the state's swamp and marshlands could be a sec- ond corn belt, a major truck-gardening area, a new citrus belt, and a rice bowl. By the early 1900s Louisiana's reclamation effort was part of a larger national endeavor.[61] Since many of these newly reclaimed lands were at or near sea level, drainage was critical. The threat of floods was always a con- cern, but landowners ignored it because they perceived the levees as safe and sound. The 1927 flood changed this perception radically and opened a new chapter in the history of Mississippi River levees. This single event changed the face of America; it was the keystone to the most comprehen- sive legislation the government had ever enacted. In the final analysis, the 1927 flood effected the entire Mississippi valley and helped lay the founda- tion for Franklin D. Roosevelt's New Deal.[62]

THE GREAT FLOOD OF 1927

> At noon the streets were dry and dusty. By 2 o'clock mules were drowning
> in the main streets faster than they could be unhitched from wagons. Before
> dark the homes and stores stood six feet deep in water.[63]

In the great flood of 1927 press accounts reported 226 crevasses; of these, 52 were major. This flood inundated 28,570 square miles.[64] The unprecedented number of crevasses precipitated renewed interest in levee construction. The problem was serious. In New Orleans the population began to demand action; they wanted a human-induced breach in the levee to lower the Mississippi's flood stage and protect the city. On April 26 Louisiana's Governor O. H. Simpson signed a proclamation giving permission to dynamite the levee at Caernarvon, south of New Orleans. The U.S. Army Corps of Engineers used 1,500 pounds of explosives to open a crevasse 3,213 feet wide that funneled 325,000 cubic feet per second from the main channel.[65] The crevasse reduced water pressure on New Orleans's levees. The city survived relatively unscathed, but the diversion flooded thousands of the city's rural neighbors.[66] In reality, the dynamited crevasse reopened one of the Mississippi's "ancient lines of discharge although the water did not follow the old bayou but flowed away from it toward the swamps on either side."[67] Although promised reparations for damages by the architects of the sanctioned dynamiting action, the promise to pay for all losses never materialized.

Contemporary observers called the flood of 1927 the "greatest peacetime disaster in our history."[68] It was the product of abnormally high rainfall over the thirty-one states and two Canadian provinces that comprise the Mississippi's drainage basin. Deluge after deluge increased the Mississippi's discharge rates. Crevasses siphoned off flow that reduced flood stage downriver. Nevertheless, the river submerged parts of seven states and breached more than 5 miles of levees. The current was so strong that homeowners tied their houses to trees to hold them fast. Many houses floated off their foundations, often lodging miles away from their original sites. It was apparent that the levees needed to be strengthened.

Approximately 800,000 individuals—a number nearly equal to the 1927 population of Washington, D.C.—fled their homes. The Red Cross provided meals for nearly 700,000 refugees for months. Tents, warehouses, schools, churches, and other shelters housed the refugees.[69] Since the flood posed "the most extensive health hazard ever experienced in America," relief agencies vaccinated and inoculated more than 500,000 people.[70] The 1927 flood, combined with the flood of 1882, submerged 57,600 square miles—an area larger than Delaware, Connecticut, Hawaii, Massachusetts, Maryland, New Hampshire, New Jersey, Rhode Island, and Vermont combined.[71] This flood transformed part of the nation and had a major cultural and political impact on the rest. For example, Herbert Hoover, secretary of commerce and chairman of a committee that handled the emergency, gained enough favorable news coverage to propel his presidential candidacy into the forefront.

The severity of the 1927 flood prompted the passage of the 1928 Flood Control Act. This comprehensive legislation began the process of locking the Mississippi River into a conduit with spillways constructed along its course to protect against severe flooding. Although the notion behind the original 1928 law has undergone many changes, chief among them is that the river cannot be contained within levees alone. Spillways and reservoirs must also be part of the management scheme. Since their construction in the early 1930s, the current assemblage of spillways has proven its worth repeatedly.

MODERN EXAMPLES OF BUILDING NEW LAND

Little data exist on the role crevasses play in adding sediments into the coastal lowlands, but scientists have documented this process in the delta.[72] Throughout the delta numerous crevasses have occurred in historic times. Cubit's Gap, Pass-a-Loutre, the Jump, Batiste Collette, Grand Bayou, Joseph Bayou, and probably many other points within the delta initially opened into the Mississippi's main channel as a crevasse. In several instances these diversion channels were dammed to prevent flow from moving through them. Designed only as temporary structures, floods destroyed these protective earthen dams/levees. For example, the Jump, south of Venice, originated as a crevasse in 1839. By 1840 the crevasse served as an active sediment conduit.[73] From the late 1860s into the 1980s this deltaic crevasse evolved into a progressive chain of distributary channels. Each channel served as a sediment conduit, building up through a series of small deltas or subdeltas into the shallow bays. These subdeltas are responsible for more than 80 percent of the new land built around the modern Balize deltaic complex.[74]

Sometime in the 1860s Cubit's Gap formed north of Head of Passes—

the apex of the Mississippi delta (see fig. 6.2). Some speculate that it "origi-nated from a cut made by the Navy through the bulkhead of a fisherman's canal to provide a boat passage to oyster."[75] Sediments moving through this outlet and deposited into Bay Rondo have created a subdeltaic complex of more than 100 square miles.[76] The gap was originally 2,430 feet wide, but enlarged to 3,000 feet. The break carried about 12 percent of the river's flow. In many ways Cubit's Gap is a deltaic anomaly, because it continues to dis-tribute sediments into the interdistributary basin.

Deltaic crevasse splays, when open, scour a channel into an interdis-tributary bay and within ten to fifteen years fill in the bay. While active, a veneer of sediments ten feet thick can blanket up to six square miles. Thus, in a relatively short period of time, the deposit will cover considerable ter-rain.[77] Between 1872 and 1896 the Pass-a-Loutre crevasse enlarged from a channel 3 feet wide to one 2,231 feet wide. From 1891 to 1921 the distributary network associated with the crevasse had built the coast twenty-nine feet into Garden Island Bay. A dam built to close the crevasse failed. Conse-quently, from 1891 to 1921 the break remained open, filling Garden Island Bay at a rate of 1,263 feet per year.[78] In the process of filling in the bay, the sediment-laden water submerged the remains of the old Spanish commu-nity of Balize, hiding it from various treasure hunters.[79]

Throughout the delta, small unnamed crevasses were responsible for creating minor subdeltas. The process of filling the bays is directly related to the delta's system of overlying crevasse splays. Every crevasse off the main channel in-fills the adjacent interdistributary bay. The process is not immediate; it is the result of repetitive flooding events through the crevasse over a period of 100 to 150 years. The end result is a new subdelta composed of thirty to fifty feet of bay fill over an area of 115 to 155 square miles.[80] This process is, in fact, the prototype of reengineering the hydrology of the lower Mississippi River to better meet the sediment- and water-starved demands of Louisiana's eroding wetlands.

CONTROLLED CREVASSING

In retrospect, an active system of artificial diversions as proposed in 1829, 1850, 1866, and 1874 would have replenished the marshes and offset the accumulated damage caused by leveeing the Mississippi River. The U.S. Geological Survey currently is proposing renourishing Louisiana's disap-

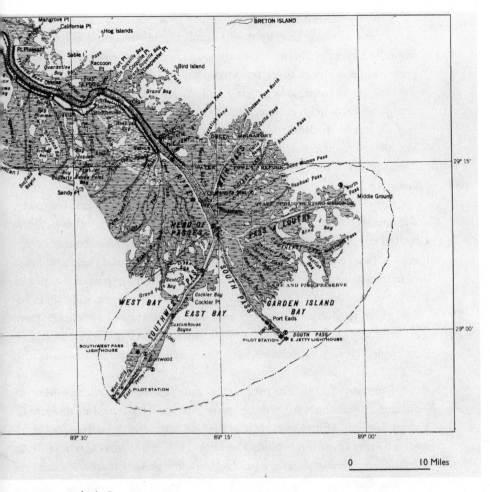

Figure 6.2. Cubit's Crevasse.

pearing wetlands.[81] However, it was not until the 1970s that Louisiana's wetland loss problem began to receive special attention from the public and
research community. Over the last two decades interest and research increased dramatically. State officials began to take action to design a plan to
reduce land loss and its associated impacts. These remedies included construction of controlled river-water diversions, regulating dredge and fill activities, advocating the beneficial use of dredge material, and promoting
land building in the Atchafalaya delta.[82]

Currently efforts are underway to design siphons that will imitate natural crevasses, while diversions will supply the freshwater to the wetlands from the Mississippi in order to restore and maintain wetlands.[83] Diversion projects at ten points along the Mississippi will introduce freshwater into the system. All of these large-scale projects are at sites of historic crevasse activity and will help create aquatic habitats for fish and wildlife resources. Not as large as the freshwater diversion projects, small-scale diversions are also like crevasses in that many are designed to promote sediment accretion. These projects augment the Mississippi's ability to replace the wetlands being lost.

Over the next twenty years the state, working with its federal partners, will create or maintain up to thirty delta crevasses. In 1998 field crews investigated seventy delta crevasses. Of these, fifteen needed maintenance dredging and two new ones would be constructed.[84] The goal of these projects is to maximize sedimentation within the interdistributary basins in order to form new deltaic lobes.[85] D. W. Roberts, a deltaic geologist, and others believe these small-scale endeavors will achieve substantial benefits.[86] They are relatively inexpensive, easy to build, and produce quick results. These controlled crevasses introduce riverborne sediments into a number of the delta's shallow bays.

The state of Louisiana proposed twenty sediment diversion outlets to prompt interdistributary sedimentation in 1992. If successful, these sediment diversion efforts will create in the next fifty years approximately eight square miles of marsh habitat. The efforts will prove particularly valuable for the estimated 250,000 ducks and 60,000 snow geese that annually migrate to these marshes.

In conjunction with funding provided under Public Law 101-646, or the Breaux-Johnson Bill (also known as the Coastal Wetlands Planning, Protection, and Restoration Act), Louisiana is working to restore, preserve, and enhance the state's eroding wetlands.[87] Current restoration techniques include marsh management, shallow-bay terracing, sediment capturing, structural shoreline protection against erosion, and emulating crevassing through diversion projects that promote sedimentation within the wetlands. Historically, the crevasse process can provide valuable information for evaluating the significance of an engineered crevasse's ability to mimic nature's delta-building processes.

The Mississippi delta projects that emulate crevasses can provide important data in the analysis of the success of small-scale diversion efforts. Since they represent a cost-effective way to maximize movement of sediment-laden waters into sediment-starved areas, these controlled crevasses should be monitored carefully. If projects are to be placed along the Mississippi's lower course, the historical sequence of crevasses along the river needs to be fully understood in order to design sediment siphons that emulate these historic events.

SUMMARY AND CONCLUSIONS

Most crevasses reported in the period 1850 to 1920 were the result of (1) paths worn across the levee, (2) poor maintenance where, in some cases, a shovel or two of dirt could have prevented the break, (3) defective rice flumes, (4) crawfish and muskrat holes, and (5) sand boils or "blow outs." Crevasses associated with these elements could, in most cases, have been stopped if riparian landowners had used proper defensive measures (sacks of sand, lumber, and repair crews) and surveillance techniques. However, nothing of the sort happened. Flooding seemed inevitable. In fact, without a system of protective levees, floods were a perpetual dilemma that early settlers learned to endure.

Inundations nourished the land, but they remained an aggravating part of living within the Mississippi River's floodplain. To neutralize these seasonal disasters, riparian landowners augmented the natural levees with engineered structures high enough to counteract floodwater or at least minimize its effects. Actually, these fabricated levees were the only practical technique available to floodplain occupants for flood protection. After the devastating 1927 flood, the U.S. Army Corps of Engineers began to construct the Mississippi's guide levees and spillways.

Today this levee network protects cities, towns, villages, farmlands, and industrial complexes. In retrospect, the levees had a dramatic impact on the general ecology of the wetlands and modified the orderly distribution of freshwater out of the river into the marsh-estuary complexes. Furthermore, they permanently altered the natural wetland processes, interlevee–basin drainage regimes, and vegetation patterns. Engineers brought about these changes through the use of levees, internal drains, and pumps. Through time, the Mississippi's levees were strengthened, eliminating overbank flooding and the systematic sediment recharge to Louisiana's subsiding coastal lowlands through crevasses. Sediment flow was effectively shut off. Natural accretion, derived from overbank flooding, was terminated. A robust system of engineered levees is responsible. Engineered to protect the population living within the river's alluvial valley, this levee system also altered the region's natural topography. Although originally considered too expensive, engineered crevasses are evolving into important components in the battle to protect and save Louisiana's wetlands.

Long-term protection, preservation, and restoration of Louisiana's wetland resource base cannot be accomplished without diverting sediment-laden water from the Mississippi River. Historically, crevasses and overbank flooding renourished the wetlands. Crevasses served as conduits to direct sediment-laden water throughout the corridor affected by the break. Sediments are critical in rebuilding the wetland. They are necessary in order to help offset sedimentation shortages partially induced by channelizing the Mississippi. At present, crevasses and overbank flooding are no longer "nourishing" the wetlands. Humankind engineered a levee system to protect against flooding. These levees successfully deprived the wetlands of valuable sediments, which directly affects the natural resource base, local communities, and public infrastructure—impacts that are real, undeniable, and to some unimaginable, but deserve real solutions. From a historical perspective, crevasses are sediment conduits; they may be one element in the solution to wetland loss.

Growing Demands of the City

Urban areas place different demands on the environment than rural areas. The intensity of land uses and the concentration of activity require adjustments in natural systems to sustain urban activity. As New Orleans's population more than quadrupled, from about 100,000 in 1850 to over 450,000 in 1930, the urbanized territory pressed outward from a small core along the natural levees near the French Quarter. By 1930 residential development stretched from the upper parish line to the lower parish line along the well-drained natural levee and back toward the lakefront.

The extension of New Orleans toward the lakefront was dependent on drainage of the swamps and marsh. Spurred by nineteenth-century beliefs that wetlands produced miasmas and federal policies that encouraged their drainage, New Orleans set out to occupy the backwater swamp situated between the city and Lake Pontchartrain.[1] Several ill-fated drainage efforts led to a more coordinated public-sector approach in the late nineteenth century. By the 1930s the Sewerage and Water Board had engineered a system to remove runoff and lower the water table within the city. With increasing property values, the private sector augmented public efforts and pushed the sphere of urban influence outward from the river-hugging settlement.

Joel Tarr has traced urban efforts to shift the burden of pollution from land to water sinks.[2] By exporting sewage, smoke, and garbage, cities extended the impact of urbanization beyond the city limits. New Orleans sent its sewage and garbage wastes downstream with the Mississippi in the nineteenth century—using a water sink to solve an environmental problem. Levees and other flood-control devices constructed to protect the city, like pollution-control projects, displaced risk to other locations.

With the construction of ever more competent levees surrounding the city, flood risks rose elsewhere. Public policy sought to provide heightened protection for the largest urban area on the lower river.[3] By the early twentieth century the river flowed through a narrow, levee-lined valley—much narrower than the natural floodplain that was several miles wide. The levees did not prevent floods, they merely kept the floodwaters from flowing down the city streets and this required a constant effort to maintain sufficient levee heights in the face of rising river stages caused by the constricted floodplain. A prominent part of the urban infrastructure in New Orleans is the levee system along the river and, more recently, along the lakefront (to keep hurricane-driven waves from inundating lakefront residential areas).

As with pollution, the levee system displaces risk, and when floodwaters rise, the areas with the strongest and highest levees escape inundation. Other areas with lesser bulwarks may endure breaches and flooding. Emergency efforts to reinforce the levees place the greatest resources on the zones of greatest value: cities. Political and economic influence also play a role in risk displacement. Urban areas and their valued real estate and greater population concentrations can and did demand shifting the risk to other locations.

Beyond land uses and economic vitality, urbanization brought with it a certain separation from the more rural areas. Cities with larger populations, lending institutions, cultural institutions, and a self-conscious press saw themselves as more sophisticated and more important than neighboring communities. This self-perception had important ramifications in managing the environment. It prompted both an unspoken cultural cost-benefit analysis in the face of disaster, as well as more obvious economic balancing of urban versus rural land uses. When the high waters of 1927 bore down on the city, civic leaders argued to sacrifice a rural parish just downstream in order to save the city. This action had profound, if short-lived, impacts on the economy and wildlife of St. Bernard Parish as Gay Gomez recounts. Similar disregard for the adjoining parish is found in Todd Shallat's discussion of the battle between the Corps of Engineers and local marsh dwellers. At issue were concerns over hurricane protection and navigation plans for New Orleans that contributed to marsh destruction. Set in a period of rising environmental concern, Shallat's chapter signifies a turning point in the relentless modification of the local environment to serve urban interests.

7

Gay M. Gomez

Perspective, Power, and Priorities

New Orleans and the Mississippi River Flood of 1927

W HO MANIPULATES nature? A simple answer would be "those with the power and resources to do so." But who are these people, and where does their motivation and empowerment lie? Furthermore, what effects do their actions have on the nonhuman world and on those who lack the power to manipulate it on the same scale? This chapter examines these questions in the context of the New Orleans area's experience in the Mississippi River Flood of 1927. It portrays a confrontation between urban and rural perspectives and priorities—a clash that echoes across time and place—and one that in 1927 pitted the interests of the city of New Orleans against those of its downriver neighbor, St. Bernard Parish.

In the mid-1920s New Orleans and St. Bernard could hardly have been more different. At the start of the decade New Orleans was the fourteenth largest city in the United States, with a population of nearly 390,000 cosmopolites. As the decade progressed the city's economy boomed along with its newly renovated port, strengthening New Orleans's linkages with the American Midwest and Latin America. Advances in drainage technology, led by Albert Baldwin Wood's invention of a powerful new screw pump, allowed the city to expand slowly northward toward Lake Pontchartrain. The progressive climate of the 1920s likely reinforced New Orleanians' superior attitudes toward their rural neighbors, people they apparently neither depended upon nor thought highly of.[1]

Among those people regarded by city dwellers as backward or even "primitive" were the residents of St. Bernard Parish, which lies east of New Orleans.[2] In contrast to its bustling, cosmopolitan neighbor St. Bernard in 1920 recorded a population of just under 5,000 people, the majority located along the narrow strip of high, fertile land along the Mississippi River's banks. The remainder of the parish's 465 square miles was an extensive wetland that stretched eastward between Lake Borgne and Breton Sound.[3] These marshes were home to several relatively isolated settlements—Yscloskey, Reggio, and Delacroix Island—where Isleños, descendants of Canary Islanders, reaped the wealth of the wetlands each season as fishermen, crabbers, waterfowl hunters, and muskrat trappers. In the 1920s muskrat trapping was the most lucrative of all. But whether a person had furs to sell, or fish, or vegetables from farms along the river, the market for St. Bernard's goods was upriver, in New Orleans.[4]

Rural hinterland and wealthy city, for all their differences, had a common concern in April 1927: the Mississippi River was rising, and an unprecedented crest was making its way toward them. Early in the month the U.S. Weather Bureau issued forecasts for a flood of "dangerous proportions" in the lower Mississippi valley.[5] Heavy rains and snowmelt had fed the river throughout its drainage basin that winter and spring. Adding to the river's burden was increased runoff caused by a decade of deforestation as farmers converted bottomlands to agricultural fields. Levees protected the fields, along with nearby towns and cities, by channeling the river's flow between earthen banks, but no spillways existed to relieve pressure on these ramparts once the water rose and began to tear at the sod and earth beneath. In places where the levees could not withstand the river's force, a crevasse or break in the levee occurred, and sediment-laden water spread across the floodplain.[6]

Crevasses were momentous events both feared and long remembered; they caused destruction of homes, loss of lives and property, and dislocation for those who survived, but they also enriched the land with nutrients and sediment, making it exceptionally fertile. In April 1927, however, people who lived within reach of the rising Mississippi River were not contemplating agricultural productivity. Some were fleeing for their lives or awaiting rescue from their roofs as major crevasses in Mississippi and Arkansas inundated Greenville, Arkansas City, and surrounding rural areas on April 21–22. Other people were reading newspaper accounts of these and earlier levee breaks along the river and its tributaries, accounts that vividly described the disaster and warned of the flood's approach downstream. Still others hurried to aid flood victims by sending boats and food and by setting up camps for the refugees.[7]

In New Orleans and neighboring St. Bernard Parish, residents feared the worst but viewed the threat from different perspectives. These perspectives illustrate each area's priorities, along with its power to deal with the flood hazard.

VIEWING THE FLOOD THREAT

Perception of the river's threat to the city was shaped largely by three forces: the New Orleans business community, the city's premier newspaper, the *Times-Picayune,* and (what residents saw all too clearly) persistent heavy rains and steadily rising water. In mid-April these forces combined to generate a wave of concern that swelled rapidly and broke at month's end, along with the St. Bernard levee.

The U.S. Weather Bureau in New Orleans began issuing flood forecasts and warnings in early April, but this information did not appear prominently in the city's newspapers. Since newspapers were the public's main source of weather information, residents during the first two weeks of April remained largely unaware of the increasing flood threat. Noting this potentially dangerous lack of newspaper coverage, the weather bureau's senior meteorologist, Isaac Monroe Cline, met with reporters handling the bureau's dispatches. They informed him that "the merchants of New Orleans had a censorship committee handling flood matters. Among the duties of this committee was . . . the suppression of the publication of flood news because such flood news would prevent country merchants from coming to New Orleans to buy goods."[8]

Cline met on April 15 with representatives of the city's Association of Commerce and its censorship committee, informing them of the magnitude of the flood upriver and urging them to cease their censorship activities. The association relented. Its plan to suppress information on the flood was intended to prevent panic and keep business booming, but, as Cline reported, their actions had carried too far. After the committee lifted its censorship the newspapers "immediately gave prominence to the flood warnings of the Weather Bureau and the sudden burst of information concerning the flood caused a panic."[9]

Indeed, the *Times-Picayune* from mid-April to the month's end proclaimed the threat of flooding in New Orleans and surrounding parishes. It published the weather bureau's predicted flood crests (including a forecast of twenty-four feet at the Carrollton gauge in New Orleans), recorded daily readings of river levels, reported on the condition of the levees, and told of efforts to raise and strengthen them with sandbags and protect them from wavewash and errant ships.[10] By April 21 major crevasses upriver became front-page news, and, despite assurances that the levees would hold, the first reports of armed "levee guards" appeared.

Levee guards were men assigned to patrol the riverfront in order to protect levees from being cut. Cutting a levee creates an artificial crevasse, which relieves pressure on nearby flood-control defenses by giving the river an outlet or spillway for its waters. As the *Times-Picayune* reported on April 25, people in the parishes downriver from New Orleans were especially concerned about their levees being cut: "Residents and Property owners along the east bank of the river in St. Bernard and Plaquemines parishes . . . arranged through voluntary subscription to double the vigilance along the east bank levees."[11] To do so they hired fifty additional levee guards, charging them with preventing unauthorized visits to the ramparts and preventing all visits at night.

One may wonder what the guards and their employers found so worthy of protection behind the levees of these rural parishes. An early twentieth-century description of the lands downriver from New Orleans provides insight: "The countryside adjacent to the river and the numerous small bayous is fertile, supporting extensive vegetable farms, sugar cane plantations, and fruit orchards. The marsh areas contiguous to Lake Borgne are unfit for cultivation, but teem with muskrats."[12]

Indeed, in addition to their homes and truck farms, the people of St. Bernard were deeply concerned about the lands that supported their economic cornerstone, the muskrat-trapping industry. These lands were brackish marshes, wet grasslands where fresh- and saltwater mixed, producing an extremely productive yet delicate ecosystem that could be easily damaged or destroyed if too much fresh- or saltwater entered the wetland.[13] When in balance, however, the brackish marshes supported luxuriant stands of

three-cornered grass (*Scirpus olneyi*), which in turn nourished large populations of the region's premier furbearer, the common muskrat (*Ondatra zibethicus rivalicius*).[14]

Large numbers of muskrats fueled a fur-trapping industry that boomed
in the 1920s as demands for fur coats soared in the United States and abroad.
By 1927 the Louisiana Department of Conservation's *Biennial Report* proudly
proclaimed that Louisiana led not only the nation but the entire continent
in fur production, surpassing even the Dominion of Canada.[15] Muskrats
comprised over 90 percent of the state's fur harvest, and in the winter of
1926–27 80 percent of these pelts came from St. Bernard Parish marshes.[16]

Muskrats meant prosperity not only for St. Bernard trappers but for the
state and its wetlands as well. The banner 1924–25 statewide harvest of 6.2
million muskrats generated $5.1 million for trappers and $62,000 for the
state via a one-cent severance tax on the pelts. Marshmen working the
129,000-acre brackish marsh around Delacroix Island supplied 800,000 furs
that year, but in the following two seasons their contribution rose to 1.79
and 2.4 million pelts in a total harvest of 3.6 and 3.0 million muskrats, respectively. With an average price of $1.20 per pelt in the 1926–27 season, St.
Bernard trappers and landowners earned approximately $2.9 million, pow-
erful testimony to the value of protecting the wildlife-rich wetlands.[17]

THE CITY'S SOLUTION

Despite the economic importance of St. Bernard's marshes and their status
as the state's premier trapping grounds, pressure began to mount in New
Orleans to cut the levee downriver. With each passing day rumors of levee
breaks and impending disaster circulated in the city, adding urgency to the
cry for action. The *Times-Picayune* attempted to quiet these rumors, but
banner headlines such as "Death Toll Mounts as Two Levees Collapse Upriver" (April 22), "'For God's Sake, Send Us Boats' Pleads [Mississippi Gov-
ernor] Murphree" (April 23), and "All in Delta Warned to Flee for Lives as
Waters Rise" (April 25) did little to allay residents' fears.[18] Spectators flocked
to the riverfront at the foot of Canal Street and at Carrollton Avenue as the
river rose to 20.9 feet on April 25. Expecting the worst, many residents prepared to ride out the flood, filling their attics with food; others left the city,
some drawing their money out of the banks as they departed.[19]

The New Orleans business community, now joined by the bankers, again mobilized its influence to quell the panic and restore confidence in the city's safety. A spillway downriver, they asserted, was the only way to ensure that New Orleans would not be overflowed. A natural crevasse in 1922 had set the precedent by creating a spillway at Poydras, twelve miles below the city. This natural levee break had siphoned enough water from the Mississippi to lower the river level two feet at New Orleans, averting a feared disaster. Despite meteorologist Cline's insistence that crevasses upstream would prevent the city from flooding by lowering the river's crest, clamor for a spillway grew louder and louder.[20]

Demands for a levee cut near Poydras reached the state capitol on April 24. Initially, Governor O. H. Simpson withheld his approval, stating that the loss of property and productivity there was not justified by the weather bureau's forecasts. When Cline revised his prediction and businessmen assured the governor that flood victims would be sheltered, fed, and indemnified for their losses, Simpson relented. After receiving approval from the U.S. War Department and the Army Corps of Engineers, the governor on April 26 issued an emergency order to cut the levee near Poydras three days later.[21]

The governor's decision to cut the levee downriver from the city, rather than upstream where the levee was already weak and in need of costly shoring, resulted from businessmen's reminders that the lands upriver were more developed, and reimbursement for damages there would be far more costly to the city. No doubt Simpson had also weighed the political ramifications of inundating the lands of wealthy and influential citizens in the sugar-producing parishes upriver from New Orleans.[22]

On April 27 the *Times-Picayune*'s headlines trumpeted news of the artificial crevasse, and with it the city's salvation: "Cut Poydras Levee Friday, Governor Orders—Flood Crisis in City Will Be Relieved." The somber tone of previous weeks had changed overnight to one of excitement, even exhilaration. As engineers prepared to dynamite the Poydras levee at Caernarvon, fourteen miles below New Orleans, residents of St. Bernard were hurried onto trucks and transported to an army supply base that would serve as their home for the next three months.[23]

The *Times-Picayune* described the exodus from St. Bernard in an April 28 story that reveals a decidedly urban perspective: "Simple souls and primitive, bittered with the loss of their homes, their gardens and their fisheries, they are, nevertheless, resigned at last to fate and secure in their faith in a rich neighbor which has turned to them in its hour of danger."[24]

Not surprisingly, trappers were the last to abandon their valuable lands on behalf of this "rich neighbor":

> But still it was in the waterlogged, swamped communities, miles from the levee, that the threat of violence lingered. Mosquito-bitten and sunburned men still spoke of the trapper's war of a few months ago and intimated that would be a Sunday school picnic to what would happen in the event of an attempt to cut the levee. They cooled, however, wavered at last and agreed to be ready to pack up and go the first thing in the morning.[25]

Why? According to Joseph "Chelito" Campo of Delacroix Island, St. Bernard residents' reliance on the New Orleans market was a powerful tool of persuasion. Campo, who was thirty-one years old at the time of the flood, recalled,

> A man came to tell us to get out. If the crevasse would not be opened then New Orleans would be lost and we all would be lost as well. [This was because the City of New Orleans was the market for the Isleños' products, and its demise would bankrupt the Isleños even if they managed to ride out the flood.] We understood.[26]

Campo, like trappers throughout south Louisiana, relied exclusively on New Orleans fur dealers to market his pelts to fur processors in New York and abroad. Dealers purchased these pelts from fur buyers, middlemen who traveled to trappers' marshland camps regularly during the December-through-February season, buying furs and selling provisions to the trappers and their families. Those marshmen who had obtained credit or loans from New Orleans merchants and bankers were usually able to pay their debts following a lucrative trapping season. A lucrative season, however, depended as much on a trapper's access to markets (via fur buyers and dealers) as on an abundance of furbearing animals in the marsh. In this city-countryside economic relationship, the city clearly held the upper hand.[27]

By noon on April 29 more than 300 refugees from the countryside had arrived at the New Orleans camp. Others sought shelter with relatives and friends. As the last of them streamed in, they met outbound city and federal officials, dignitaries, and newspaper reporters, all hurrying to the Caernarvon levee to witness the momentous event. Waiting for the reporters was

St. Bernard Parish sheriff L. A. Meraux, who made clear his perspective on the impending crevasse:

> We're letting 'em do it because we can't stop em. . . . You can't fight the Government. I have a hell of a time trying to get my people to see that. . . . They wanted to go to the levee first with their women and children and their weapons, and tell the State of Louisiana to come ahead and cut the levee—but it would be cut over their dead bodies first. We managed to talk them out of that for their own good. But this ruins us. . . .
>
> Here's the parish of St. Bernard. Supplies more furs to the world than Canada and Russia put together. The trapping will be ruined for from three to five years. We've got the best oysters in the world. The oyster trade will be ruined, and the shrimpers and the fishermen. The truck farms are gone. The cattle and the canning industries will be wrecked. And we haven't got a line in writing for any kind of a guarantee that we're going to get anything back out of all we lose.[28]

At noon levee engineers detonated the first charge of dynamite; additional blasts over the next four hours initiated the Caernarvon crevasse, a series of levee punctures that siphoned the Mississippi's waters across St. Bernard's wetlands to Lake Borgne and the Gulf of Mexico (see fig. 7.1). Standing at the edge of the first gap in the levee, Sheriff Meraux proclaimed to reporters, "Gentlemen, you have seen today the public execution of this parish."[29]

Blasts associated with the "execution" continued for an additional ten days, at last widening the crevasse to 2,800 feet. As 175,000 cubic feet of water per second flowed into St. Bernard and Upper Plaquemines Parishes, New Orleanians watched the river level slowly fall to 20.3 feet on the Carrollton gauge. News of devastating crevasses in Louisiana and Mississippi continued to dominate the front pages of the *Times-Picayune,* but its previously strained assurances that the city's levees would hold now rang with confidence. When meteorologist Cline declared on May 6 that the river's crest would not exceed the twenty-one-foot level reached on April 25, New Orleans residents breathed a sigh of relief and turned their attention to other matters, including compensating the refugees for their losses.[30]

St. Bernard trappers and wetland landowners, however, were not waiting passively for reparations. They were back in the marsh, fighting to prevent the complete disaster Sheriff Meraux had predicted. Their efforts illustrate a rural people's vigorous response to a flood released upon them by a wealthy and powerful urban neighbor.

Figure 7.1. Water pouring through the Caernarvon crevasse began flooding St. Bernard Parish (right) on April 29, 1927. *(Courtesy National Archives, photo no. 77-MRF-173.)*

THE RESPONSE IN ST. BERNARD

When trappers realized they could not prevent the inundation, they immediately turned their attention to saving a "seed stock" of muskrats to repopulate the famed trapping grounds once the water subsided and salinity increased, allowing the animals' food grasses to regenerate. In this way they hoped to hasten the fur industry's recovery, restoring an important source of income for the parish, the state, and especially for the individuals involved in the trade.[31]

To save muskrats from the flood, trappers, landowners, and Louisiana Department of Conservation personnel quickly mounted a very unusual rescue effort. First, the Department of Conservation prohibited the sale and purchase of furs from the inundated areas, thereby protecting the animals from a frenzied early harvest.[32] Next, seasoned marshmen realized action was needed at the flood scene, for although the nocturnal muskrat forages in aquatic habitats it can survive neither long periods of immersion nor prolonged exposure to the sun.[33] As three feet of water swept through human houses and easily tore apart the muskrats' grass dwelling mounds, trappers and wetland landowners answered the call to action. Traversing the flooded marshes in pirogues and canoes, the rescuers found

> muskrats, young and old, swimming with the swift currents. Some were energetically making their way somewhere and with evident unconcern and when we neared them, bent on rescue, they would dive beneath the surface with a resounding slap of the tail and defy our good intentions. Others, sensing possible rescue, would make straight for our boats where they would climb paddles or suffer themselves to be lifted from the water with long-poled nets, and some even permitted us to take them aboard with bare hands; . . . when placed on the bottom of the canoe [they] would lay supine from exhaustion.[34]

During the flood's first days, trappers and conservation agents transported thousands of these "refugee 'rats" to higher ground, but much work remained in the marshes near Delacroix Island. As the Department of Conservation reported, "Trappers worked practically 26 hours a day to succor the very animals they pursued with their steel traps every winter and, for the first time in history, life rafts for muskrats were hastily constructed and floated on the vast yellow inland sea" (232).

Trappers had conceived the idea of liferafts after observing surviving animals riding the flood on anything that would float, including platforms of grass and reeds the animals had gathered. In order to provide additional platforms, the rescuers first built simple wooden rafts and launched them, along with "skiffs, fish-cars, pirogues, rowboats, ordinary packing crates— anything that would float and carry a cargo of Spanish moss, tree branches, and grasses" (238). Anchored in the flooded marshes, they provided temporary relief while trappers hurried to build more substantial craft.

Manuel Molero, owner of Delacroix's Acme Land and Fur Company, supplied lumber for the rafts, while Harvey Roberts, a trapper and marshland owner, designed the liferaft that proved most effective. It consisted of

> four 1x6 cypress planks, two inches thick and placed four inches apart, [which] were nailed to three 2x4 inch crosspieces. A foot-wide plank was attached to each side of the raft and held in place by another crosspiece laid across the exact middle of the top. Over this centre crosspiece was heaped a great armful of three-cornered rush, the favorite food of the muskrats of this section. (237–38)

Figure 7.2. A rescued muskrat seeks "higher ground" on the shoulder of a conservation agent in St. Bernard Parish. *(Louisiana Department of Conservation, Eighth Biennial Report, 241; courtesy Special Collections Division, Tulane University Library.)*

These improved rafts provided muskrats with a temporary haven and food source, allowing the animals to feed, take shelter, and give birth as they recuperated from a long immersion or prepared for another. During the 108-day inundation, an estimated 1,200 rafts floated on the floodwaters, and the Department of Conservation believes they succeeded in saving some of the muskrats that took refuge on them (237–39) (see fig. 7.2).

While drowning and exposure killed an estimated one to two million muskrats, the survivors, along with animals brought in as part of a restocking effort, flourished once three-cornered grass began to grow again in profusion. A mere two and one-half years after the Caernarvon crevasse, trappers were back in the marshes, harvesting 900,000 million muskrats in the winter of 1929–30 and 1.1 million muskrats the following season. Although the harvest never again reached preflood totals, the trapping industry in St. Bernard Parish had recovered, along with the marsh that supported it.[35]

THE PRICE OF VICTORY

New Orleans bankers, businessmen, and city officials had pledged to fully compensate residents of St. Bernard and Upper Plaquemines Parishes for their losses in the flood of 1927. These losses included destroyed and damaged homes, ruined crops, and lost income from wildlife harvest. Faced with a total of $35 million in claims, however, the city's Reparations Committee quickly enacted rules that reduced the settlement to $2.9 million. One of these rules decreed that claimants would only be compensated for personal property lost in the flood, and since muskrats were deemed property of the state trappers would not be reimbursed for their lost harvests. Only Manuel Molero of Acme Land and Fur Company was successful in obtaining compensation. Molero hired a lawyer to defend his claim, and his company received the committee's largest settlement, $1.5 million.[36]

For the New Orleans business community, $2.9 million seemed a small price to pay for the city's safety and continued prosperity. Not surprisingly, many St. Bernard residents disagreed, and resentment against the "rich neighbor" that had singled out their lands for sacrifice lingered for decades.[37] In the years that followed the Caernarvon crevasse, this resentment received political expression as Governor Huey P. Long and his successors shifted the balance of power from city to countryside. Residents of St. Bernard and Plaquemines Parishes were willing allies of Long; the two parishes shared a congressional seat with their upriver neighbor, which they consistently outvoted.[38]

In 1927, however, urban perspectives and priorities prevailed easily over their rural counterparts as New Orleanians successfully deflected a perceived flood hazard to the adjoining parish. While the trappers of St. Bernard succeeded in restoring their livelihood, the 1927 flood was a precursor of later confrontations between urban and rural perspectives, priorities, and power —confrontations from which St. Bernard Parish's wildlife-rich marshes, and the culture and economy they sustained, have yet to recover.[39]

In the Wake of Hurricane Betsy

A HURRICANE IS a tropical furnace that draws hot air inward and up. Water vapor condenses, releasing energy: enough power to swirl one billion tons of moisture at 1,000 feet per second, enough to run a 300-trillion-horsepower motor or light and heat the United States for a year. The fury spins with a force indifferent to civilization, but damage results from human factors. Its power to kill and destroy is a variable, closely dependent on human caution or daring, on sound or shoddy construction, on the lure of seaside housing with an open view of the beachfront and sprawling urbanization toward ever more marginal land.[1]

A hurricane, in short, is a thing of nature, but the disaster it causes is not. "A lot of us argue that increased urbanization and wealth are responsible," says Chris Tucker of Emergency Preparedness Canada. "Look at Los Angeles and Mexico City and Vancouver with their earthquake risk, or Miami and Galveston with their hurricanes. . . . People don't think about where or how they build. They just ignore the natural hazards."[2] People don't think much about wood housing in a tinderbox grassland or condos anchored to dunes that move with the migrating coast. Even in the metropolis leveed between a coil of the Mississippi and hurricane swells from the Gulf, suburbia fills the floodplain. But "the City That Care Forgot" has always flirted with devastation. Built in a swamp, rebuilt in 1722 after a fifteen-hour tempest destroyed houses and sank warships, unscathed by the murderous flood of record in 1927, saved by a massive diversion through the untested Bonnet Carré spillway in 1937, battered but spared again when Hurricanes Flossy (1956), Carla (1961), and Hilda (1964) ravaged and killed nearby, New Orleans has long survived on the hope that ever more massive construction can sustain industrial growth. "We have spent hundreds of millions of dollars to protect ourselves from water," said Louisiana governor John J. McKeithen in 1965, explaining New Orleans on the eve of its greatest disaster. "We have cut the Mississippi in many places so the water can get faster and quicker to the gulf. We have built levees up and down the Mississippi. . . . We feel like now we are almost completely protected."[3]

Blessed with oil and gas and fortified at the tip of the world's most sophisticated levee and spillways system, New Orleans seemed "almost completely protected" in the Kennedy-Johnson years of robust urbanization, back when Americans had towering faith in monumental construction and a killer named Hurricane Betsy, on September 9, 1965, breached the levees of Lake Pontchartrain.

BETSY'S TOLL

Large but shallow, a 635-square-mile blue-green sheet of tideswept estuary with an average depth no deeper than a conventional basement, Lake Pontchartrain rolls hurricane swells from the Gulf of Mexico into southeast Louisiana about every decade or so. Conquistadors searching for golden cities recorded the violence of tree-ripping winds on September 19, 1559. Since that time 172 hurricanes have raked coastal Louisiana. Thirty-eight have reached New Orleans via Lake Pontchartrain. In 1893, to cite the most deadly example, a day-long barrage of thirty-foot swells overtopped coastal levees and flattened a fishing village south of New Orleans, killing 1,500. In 1915 a powerful storm took out a rail bridge and destroyed a lakeside farm colony, killing 250. In 1940 a week of punishing rain marooned thousands of people but killed only eight.

More dangerous storms caused more devastation, but science and engineering fought back. Better roads, stronger bridges, home radios, weather radar, rescue helicopters, satellite tracking: each advance in the technology of evacuation was a victory against what had always been the most lethal of nature's elements, the element of surprise.[4]

The downside of those life-saving innovations was a false security that exposed the lakeshore suburbs to waves surging in from the Gulf. Technology was seldom an unmixed blessing, and certainly not in Louisiana where the system built to funnel and deepen the Mississippi also prevented the river's silt from rebuilding a fraying delta, where dredging and oil drilling aggravated the sinking and deterioration that inched New Orleans toward the Gulf. "We have defused problems by diffusing them," wrote Edward Tenner of Princeton in a recent critique of technological overdependence. "We have exchanged risk to human life for greater exposure to property damage, and then distributed the cost of that damage over space and over time. We have assumed an increasing burden of vigilance along with our protection. Technology is again taking its revenge by converting catastrophic events into chronic conditions—even as natural catastrophes persist."[5]

Vigilance against the routine that obscured the danger of the catastrophic made a record-setting disaster of Hurricane Betsy. Killing surprisingly few but causing remarkable damage, the storm changed little about the way engineers defended New Orleans. It showed, however, that good engineering could have bad consequences, that much of what humanity had built for the safety of civilization made life in an unstable city less stable than ever before.

Hurricane Betsy began innocently as a "nice round circular storm" off the coast of Venezuela. On August 27, 1965, the hurricane gained strength and speed as it swept north toward Georgia, looped back for a day over Nassau, and then, looping again, pounded the Florida coast. On the morning of September 7 a surge off Biscayne Bay invaded Miami, washing eels through ornate hotels. In a day Florida lost 90 percent of its avocado crop, half of its lime harvest, and the windows and shutters and roofs from 4,300 houses. Seven people died. It was "the meanest wind ever to sweep out of the Caribbean," warned a teletype from the U.S. Coast Guard. Mean *and* hard to predict. Curving west on September 8, Betsy entered the Gulf with a bead on Galveston Bay.[6]

"Well, those Texans will just handle that storm just like that," the Louisiana governor said. Not yet a year and four months in office, John McKeithen had already weathered powerful Hurricane Hilda, and he doubted that misfortune would strike so quickly again. "It just isn't quite fair," he protested when Betsy seemed to be veering north. It was Thursday, September 9, and McKeithen was with family upstate for a high school football game. At 1:30 P.M. the telephone rang with an update. "This thing," said a staffer in Baton Rogue, "is headed for our state."[7]

At dusk a wave thrown off the Gulf crushed Grand Isle. Breaking pipelines, stripping trees, killing livestock, and floating houses onto the levees or tossing them into the marsh, the great hurricane launched a surge that sheeted across 2.5 million acres. The flooded port of Venice disappeared from satellite photos. Wind gusts up to 170 MPH twisted oil rigs and ripped anchor cables. Eleven shipwrecks blocked thirty miles of the swollen Mississippi. There were also forty-six "sunken objects" and more than one hundred smashed or grounded barges. One barge tanker went down with enough liquid chlorine to kill a city of 40,000 if exposed to the air.

Drifting north with a forward speed of 20 MPH, Betsy hit downtown New Orleans soon after nightfall on the ninth of September. Telephone poles snapped. Glass and twisted steel pelted Canal Street. Flooding consumed Claiborne Avenue and Gentilly. Looters used scuba gear. Just before midnight a green wall of foam off Lake Pontchartrain entered the Industrial Canal and raced toward the Mississippi. Rising to twelve feet as it swept warehouses and railyards in the lower Ninth Ward of New Orleans, carrying corpses and cargo and cars, and finding gaps in the unfinished levees between the port's inner harbor and the Parish of St. Bernard the flood left pools almost as deep as the lake in suburban Arabi and Chalmette.[8]

"I felt something cold, looked down and there I was with water in my lap," said a man who had fallen asleep at the television before the tide off Pontchartrain flooded into Chalmette. "God it was like one giant swimming pool as far as the eye could see. There were people I knew—women, children, screaming, praying. . . . A women who lives down the block floated past me, with her two children beside her." A father of eight, strength ebbing, struggled against the Pontchartrain flood with five young children. "I couldn't do it," he grieved to a *Newsweek* reporter. "I had to let two of them go."[9]

Six thousand houses sustained serious damage below the Port of New Orleans. Twenty thousand people barely escaped with their lives. "People waded shoulder-high in water with babies on their heads," recalled a resident of the hard-hit St. Bernard Parish Carolyn Park subdivision.[10] One man escaped by punching a hole in his ceiling, breaking both hands and his back. Another was saved by a plastic raincoat. "The winds took me," said Eddie Ste. Marie, badly shaken. "I don't know how high I went but I knew I

Figure 8.1. Helicopter hovers over Betsy-ravaged New Orleans, September 10, 1965. *(Courtesy National Oceanic and Atmospheric Administration.)*

was turning and when I come back down it [the raincoat] formed a parachute. . . . I saw myself dying. I knew I was going to die."[11]

It could have been worse. In 1957 a hurricane weaker than Betsy had killed 556 when victims near St. Charles refused to evacuate. Eight years later an emergency plan was in place. The 4th Army stood ready with 700 crates of sanitation equipment, 1,000 drums of water, 8,571 mattress, 21,430 blankets, and 117,000 gas masks. The Red Cross served 1 million cups of coffee and 500 million plates of food. Rescue boats rushed to the hurricane zone from as far away as Shreveport. With twenty Coast Guard helicopters and radar support from a U.S. destroyer and Air Force reconnaissance planes, the relief operation was unprecedented. Governor McKeithen called it "a great, great example of Americanism." Even so, the catastrophe was unlike anything the delta had seen in peacetime. Injuring 17,600 people and killing 81, Hurricane Betsy, said McKeithen, was "the worst disaster in the state since the Civil War."[12] The insurance industry reeled in shock. Betsy, said an insurance spokesman, was "the worse natural disaster in [the history of] America—greater even than the San Francisco Earthquake and the Chicago Fire combined."[13] Damage estimates ran as high as $2.4 billion—more costly than any storm on record at the time, the inflation-adjusted equivalent of $12 billion today.

The dollar devastation was large but not large enough to shake a city's sense of itself as a hub of technological progress. In 1966 the recovering city took in $1 million a day from the petroleum industry, $50 million a year from business conventions, $175 million from tourism in the French Quarter, $190 million in payroll from 27,000 new NASA rocket assembly jobs. No matter that housing foundations were sinking, or the oyster harvest was down, or the marsh below the city was fraying like rotten cloth. New Orleans was making a comeback. "Positive thinking has become infectious in the new boom city of the Southland," said *U.S. News and World Report.*[14]

New Orleans was riding too high for talk of the next disaster. Prosperity was fueling itself.

A HURRICANE "SUPERHIGHWAY"

"See that?" Fisherman Frank "Big Kenny" Campo guns his launch toward the ruin of his father's marina in the shallows east of New Orleans. "Over there, that was Shell Beach. Dry land. That light over there was the shore."[15]

Dark and barrel-chested, an Isleño whose peasant foreparents emigrated from the Canary Islands to colonize Louisiana and fight the British for Spain, Kenny, age 50, runs a shanty called Blackie's Marina from a slip off a freight canal. No fax or internet listing. No shells. No beach. Wakes slap a small lagoon dank with the rot of swamp grass. Vines and climbing ivy strangle each other for light.

Figure 8.2. The eroding Mississippi River Gulf Outlet near Shell Beach on the western shore of Lake Borgne, about 1996. *(Courtesy New Orleans District, U.S. Army Corps of Engineers.)*

"It's been, what, thirty-five years since they built the canal? We've sunk thirty inches," Kenny continues.

He kills the outboard and glides. A gray man on a timber dock waves a hose at a styrofoam ice chest. "I'll tell you what's wrong with the marsh," says Blackie Campo. Father and partner to Kenny and grandfather to a hulk of a youth the family calls "Little Kenny," Blackie, age eighty, remembers green flats of scrub vegetation. Shell roads with lumbering oxen. Palm roofs on tarpaper huts. He remembers Bayou Terre aux Bouefs ("land of the oxen") in the Parish of St. Bernard where Isleños for 200 years had guarded their isolation. And Blackie darkly recalls the last week of April 1927 when the U.S. Army Corps of Engineers, with presidential approval and thirty-nine tons of explosives, blasted the record flood into St. Bernard. It was "the public execution of the parish," said a marsh politician, and it haunts the grasslands still. "Make a list of everything the Corps does around here," Blackie advises. "Everything on the list . . . *that's* what's wrong with the marsh."[16]

Once the parish, rich with muskrats, was a fur empire second to none. Blackie remembers when Louisiana exported more fur than all of Canada or Russia. When a silk-lined coat of Genuine Natural Muskrat was, said Sears and Roebuck in 1927, "a remarkable value" at $199.[17] Living wet and eating from tins, a trapper could net a profit—maybe $4,000 a year. He could also net redfish and trout. "We'd trap in the winter, fish in the sum-mer," says Kenny with a glance at Blackie. "Shell Beach had fifty families at one time. We had a train to the French Market [in New Orleans]. The tracks went into the water. Right to the [fishing] schooners. How big were those schooners, daddy? Fifty feet? We'd fish Lake Borgne, mostly. Fish sold by the string, by 'the hand' we used to call it. Croaker, flounder, sheepshead, mack-erel. . . . Best fish in the world. We never kept anything under two feet."

That was before Mister Go. Completed in 1963 at a cost of $95 million, the Mississippi River Gulf Outlet (called MRGO or "Mister Go") sliced forty miles from the treacherous trip through the delta's meandering passes. Fog and mud bars had plagued navigation through the mouth of the river since 1719 when the French first sounded a channel for warships. In 1838 the U.S. Corps of Engineers scraped at the mud with an endless-chain bucket dredge. The chain broke. Spare parts went down in a shipwreck. After two years of bad luck and mechanical frustration, the captain in charge was convinced that a ship canal was the surest solution, that "any sum, however large, would not be more than commensurate with its [the canal's] importance."[18] A century later the project was "vital," said Senator Russell B. Long, "to the maintenance of the position we hold in the free world today."[19] MRGO, a boon to the Port of New Orleans, was Louisiana's response to grain-trade competition from the new St. Lawrence Seaway. It was also a saline cut through the core of Iberian culture in the Parish of St. Bernard.

Already that parish below New Orleans was pooling with open water. Barely two feet above sea level in 1960, the parish, heavy with mud, had been dropping a tenth of an inch each year. Now brine pushed inland by

Figure 8.3. Houses swept onto the levee, St. Bernard Parish, September 1965.
(Courtesy New Orleans District, U.S. Army Corps of Engineers.)

storms ate chunks from the floating reedflats. Shell Beach gave way to Lake
Borgne.[20]

By 1960 five or six houses on stilts had relocated to a paper-thin puzzle
of swamp grass. Storms sped across shallow water faster than ever before.
"Betsy moved quick," said Blackie. Too quick, said those who suspected the
engineers were somehow involved. It had been thirty-eight years since the
Corps had blasted the levee. Had the Corps been at it again? "They give us
too much credit," says Harley Winer, a Corps hydrologist. "Do they think
we were out there with dynamite in 100 MPH winds? It's ridiculous."[21]

But rumors persist. "Yes sir, they had a plan [to blow the parish levees].
That's what my cousin told me two days *before* the flood," claims Blackie.

When the storm hit two days later, Blackie and Kenny were rushing boats to the cover of swamp. "We were towing skiffs into the forest," Blackie explains. "I pulled hard against the current. The [MRGO] channel was high [and] the stronger it blew, the higher it got. That's how the parish flooded."

Mister Go, said the *Times-Picayune,* had become a "hurricane super-highway."[22] That's the conventional wisdom in New Orleans, the gospel truth in St. Bernard. And yet it's absurd. "Let's just call it a popular misconception," says Robert J. Guizerix, chief of structural engineering at the New Orleans Corps. An earnest man in a spotless office, he, too, was a witness to Betsy, and this New Orleans native has spent a career making sure levees don't fail again. No 500-foot canal could divert a tidal surge, says Guizerix. Nothing as big as Betsy could be captured by anything humanmade. "A hurricane can be 200 miles across. A surge can be 100, maybe 150 miles wide. It's not going to even notice a tiny cut. It wouldn't even know it was there."[23]

Blackie smiles. Rocking back on a porch tacked to a house that was built from the wreckage of other houses, a house that the engineers "just picked up and moved" to make room for Mister Go, Blackie has already heard too many government experts who dismiss what a fisherman knows. He remembers the Corps telling Congress that Mister Go could not have flooded the parish because a 1915 hurricane with the same approach had done similar harm. And he remembers the post-Betsy testimony of a Brit named Ian Collins. "He was some kind of scientist from England. A red-haired limey! This guy's going to tell *me* which way the water was flowing? I was there." When the experts denied a link between the channel and hurricane flooding, St. Bernard and six other plaintiffs sued for flood compensation. "The inadequacy of the [levee] system was fully known, fully appreciated, and the consequences were understood," said a parish spokesman.[24] After a decade of charges and countercharges, a 5th District judge tossed the matter from court.

Defeated by government science, the campaign against Mister Go shifted to a war of attrition on murkier legal ground. Congress helped. With President Richard Nixon's signature on January 1, 1970, the National Environmental Policy Act (NEPA) became the first and most far-reaching legislative attempt to "prevent or eliminate damage to the environment and biosphere." A Magna Carta for environmental protection, NEPA made federal builders confront the dark side of engineering—the dislocation, the danger to wild species, the risk of planning gone wrong. The law cut to the bone of Corps economics. By questioning the way the agency had been allowed to define the "benefits" of river improvement, it invited a frank disclosure of what the critics called "social cost."[25]

NEPA mandated public hearings—a chance for officials in St. Bernard to protest and stall construction at every critical stage. In 1972 the target became the government's plan to cargo traffic to a gleaming new mechanized port. Centroport U.S.A., as the project came to be called, would be the nation's biggest ship lock, a Corps-built Port of New Orleans Mister Go extension that would bypass the congested and antiquated urban harbor. Boosters coveted the open space at a small landing called Violet in St. Bernard Parish. Pete Savoye of the parish sportsmen's league, now a retired carpenter from Chalmette, remembers the local talk: "Hey, we're going to have waterfront property! We're going to have warehouses and industry!"[26] Supertanker and container ships would enrich the parish as they crossed back and forth from river to seaway. Rising like an industrial phoenix about ten miles southeast of Canal Street, the rebuilt landing at Violet would be the Rotterdam of the South.[27]

So said the Corps, the New Orleans dock board, and Louisiana's powerful Tidewater Development Association. Land developers even mailed out a flyer that promised an annual payroll of $80 million from 10,000 industrial jobs. Negotiations quickly unraveled, however. Who would make important decisions about issues affecting the parish? What about marsh degradation? What about the hurricane threat? Savoye recalls, "You'd have to have been an idiot to believe some of the things they were saying. You'd have to have been a fool. The politicians said these canals would bring us all kinds of goodies. The parish never got a penny. What we got was erosion and floods."[28]

Corps ecologist Sue Hawes was prepared for ideological conflict but unprepared for the rancor of Centroport NEPA hearings. On November 29, 1972, for example, a seething crowd of about 700 ambushed one of the nation's most powerful politicians. Congressman F. Edward Hebert, a longtime Mister Go supporter, said the economy would simply "perish" if the port did not expand. Appealing to patriotism, he asked his constituents to "accept and embrace the common good." Appealing to wallets, he promised to "turn the [the Mister Go spur canal] project over to St. Bernard Parish for development after its completion. This means—hear me well— this means that the New Orleans Dock Board . . . will have nothing to do with any control over any of the land."[29] But the parish insurgents had heard it all too often. Chanting and jeering and parading through the room with placards, the protesters drove Hebert off the auditorium stage. Years later Hebert claimed he had never been so rudely treated. Quite a statement. The Isleños of St. Bernard had once defended the bayous with shotguns. Politics had seldom been calm.

Back on River Road in the old army barracks converted to a government compound, the Corps responded with outrage. "We'll go down there and build the project with tanks and guns if we have to!" said Colonel Richard L. Hunt.[30] Politically, however, the bullish era of Army-directed construction was already dead. Soon the Tennessee-Tombigbee Waterway was halted by court injunction; so was the Cross Florida Barge Canal, New Melonis Dam on the Stanislaus, Oakley Dam on the Sangamon, and the Baldwin Channel proposal to float supertankers into California's Central Valley. In Missouri, after a series of NEPA hearings led to embarrassing headlines, the Corps stepped back from its plan for high levees north of St. Louis. In Texas the voters rejected a plan to make Dallas–Fort Worth a seaport. In Arkansas a federal judge blocked the draining and channelization of Cache River swampland. In Idaho the state wildlife agency borrowed the language of NEPA to protest fish kills below Dworshak, the Corps's tallest dam.[31]

Water pollution was another concern. Ralph Nader's *Water Wasteland* (1971) shocked readers with the disclosure that Chesapeake Bay was dying, that bacteria in the Hudson River was 170 times the swimmable limit, that nine out of ten FDA-tested swordfish had mercury levels unsafe for human consumption. By 1972 a third of the nation's shellfishing beds had been closed due to pollution. The annual commercial shrimp harvest, 6.3 million pounds in 1936, had nosedived to 10,000 pounds.[32]

With the 1972 Federal Water Pollution Control Act—the Clean Water Act, as it came to be called—Congress responded in the same grandiose way it had often dealt with any waterborne crisis: it gave the Corps a vast and ill-defined regulatory jurisdiction, a quagmire as it turned out. The act expanded the Corps's authority to prevent mud and debris from interfering with navigation. Under section 404 of the statute, the Corps was required to issue or deny a permit for all dredging or construction in waterways or wetlands.[33] In Louisiana that meant the builder was now a sheriff. Engineers were suddenly the arbiters of land schemes and zoning disputes across 85 percent of the state. Retired colonel L. Kent Brown, a former New Orleans commander, thought the law was unfair. "Congress puts engineers out there to make controversial decisions. The Clean Water Act was supposed to be about water quality, but we used it to tell people they can't develop their property. Politicians liked it that way. They wanted to campaign against big bad government. They expected the Corps to protect the wetlands but Congress wouldn't pass the laws."[34]

As engineers struggled to interpret their ambiguous mandate, Mister Go consumed the marsh. Rapidly eroding, the original 650-foot-wide channel had destroyed 3,000 acres of marsh by the mid-1970s. Hurting fish and shellfish, it leaked pollution and salt. The U.S. Wildlife Service feared that a freightway off to Violet would destroy another 5,000 acres and "severely impair" the parish's effort to slow its slide to the Gulf.[35] In April 1976 the Chief of Engineers in Washington signed off on the Violet project. The governor of Georgia weighed in on the side of the parish. If elected to the presidency, said Jimmy Carter, he would cleave the pork alliance between Congress and the Port of New Orleans. No extension to Mister Go would bisect the pasture at Violet. On April 18, 1977, a press release from the Carter White House killed the Centroport concept, calling blandly for "further study."[36]

It had been twenty-one years since Congress first approved a new ship lock for New Orleans. Another twenty would pass before a modern lock was again under NEPA review.

A CONCRETE SOLUTION

"Russ, you know I've a couple of wars on hand," President Lyndon Baines Johnson told the son and political heir of Governor Huey Long.[37]

Louisiana senator Russell Bileau Long had been patched through to the White House at 1:30 P.M. on Friday, September 10, 1965. A teenager when his larger-than-life father had died from a gunshot in the Baton Rouge statehouse, he was now, at age 43, the Senate's majority whip. "Mr. President," said Long from his Senate office, "you know, next to the Great Lakes, Lake Pontchartrain is the biggest lake in America. You know how it stands now? Dry. Just like that, water in a wash basin. The forces of nature took it and just poured it on New Orleans." He paused, milking the moment. "When neighbors are sick and you visit them, they appreciate it."

"I'll certainly send my best man," the president offered.

"We don't want your best man, we want the head man," Long insisted. "Mr. President, we want you."

Six hours later the president was stepping from Air Force One onto the tarmac at New Orleans's Moisant Field. "I am here," the president said, "because I want to see with my own eyes what the unhappy alliance of wind and water have done to this land and its people."[38] Wind and water and *humanity*, the president might have added, because the storm had landed hard where an influx of Texas investors, Lady Bird Johnson among them, planned to levee off 32,000 acres for 250,000 people in a new suburb and industrial park. New Orleans East, as it was called, would need Corps flood protection; so would the northshore suburbs, Jefferson Parish, the Port of New Orleans, the Port of Venice, Morgan City, and a dozen or more other storm-battered sites. On October 27, 1965, Congress approved $250 million for Louisiana projects, including a $56-million downpayment on levees and storm barriers for Lake Pontchartrain.[39]

Expecting a Betsy, the Corps had published a study with a Pontchartrain hurricane plan: levees, drains, a ship lock, steel and concrete flood walls, and two gated control structures. The centerpiece would be a fortresslike barrier in the Rigolets pass where waves entered the lake. A Dutch innovation adapted by the British after a North Sea storm killed 300 near London in 1953, storm barriers with mechanical gates could instantly close off a port. The Louisiana version was to be hinged like a seawall with sixteen rotating doors. The barrier would stay open for the ebb and flow of the Gulf in normal weather. It would shut during dangerous storms. Fish biologists worried that a dead zone behind the gates might disrupt lake circulation, but in 1974 the Corps's environmental impact statement (EIS) minimized the threat to marine life. Environmentalists scoffed.[40]

Corps watchers soon understood more was at stake than marine life. Luke Fontana of New Orleans, an attorney who had crabbed and duck-hunted the black lagoons now slated for subdivisions, led a crusade linking hurricane engineering to tax-supported "land enhancement" schemes. Lady Bird's New Orleans East, for example, would have its own triangle of levees. New levees would also reclaim some 5,000 acres for suburbs in the crab-rich La Branche wetlands, in the cane and tupelo tidewater of Jefferson Parish, and in the northshore Eden Isle subdivision, where developers had landfilled the bayous only to dredge out recreational canals. Critics feared a "piracy" that would "lead to the collapse of the Pontchartrain basin as a viable system." It was "pure pork-barrel," said Fontana. Builders would reap "windfall profits" and, worse, they were trashing the public domain.[41]

With backing from northshore shrimpers and crabbers and heavyweight Jim Tripp, chief counsel for the Environmental Defense Fund in New York, Fontana filed for a NEPA injunction in the U.S. 5th District Court. What followed was high theatrics: fraud accusations, an attack on the drinking habits of an expert witness, documents photographed with a hidden camera, a red-faced Fontana swearing and slamming the door as he stomped from the judge's chamber, rumors repackaged as news in a *Times-Picayune* exposé. The Corps appeared incompetent when a government hydrologist disputed another's statistics. The Corps appeared corrupt when the plaintiffs produced an early draft of the 1974 EIS. A Corps ecologist had wishfully concluded that storm barriers "should not" interfere with the food chain. But the chief of the engineering division was not taking any chances: "should not" became a definite "will not" in the final report. That settled it. On December 30, 1977, Judge Charles Schwartz Jr. denounced the "legally inadequate" science and suspended the twelve-year-old barrier plan.[42]

Chastised and sent back to the lake with a team of Louisiana State University researchers, the Corps, by 1984, had eliminated hurricane barriers from the project design. It no longer mattered. In seventeen years the estimated expense of the project had bloated from $85 million to $924 million. The General Accounting Office doubted the barriers could be built even for that. Massive structures of concrete were no longer cost-effective. Dirt was less expensive. The Corps compensated with a great wall of earthen embankments as tall as seventeen feet.[43]

Time plays with the memory of combatants who cling to parts of the puzzle, recalling disparate things. Lake defender Luke Fontana remembers the Corps's attempt to cloud the case in a blizzard of information. Ecologist Sue Hawes remembers "bad science." Joseph Towers, a Corps attorney, remembers "an embarrassing situation." Engineer Robert Guizerix remembers mostly the sound of wind sucking water as Betsy "lifted Pontchartrain [over New Orleans] and dropped the lake like a bomb." There are few historical files in the engineer's office. He can't even recall the disputes over levees for land enhancement. "Their scientists said one thing. Ours said another." That's what Guizerix remembers. All he can say for sure is that barriers guarding the lake are the better engineering solution. Tall levees are more expensive. More expensive and not as safe.[44]

"The next Betsy—that's all it will take," says Guizerix. "When the next big hurricane comes, and it will, people will be wanting the barriers." Meanwhile, he won't be tossing out any plans.

LOSING LOUISIANA

Memory fades in the wreckage of recent disasters. Four years after Betsy did or did not ride Mister Go, 200-MPH Camille, the strongest hurricane on record, sent a twenty-four-foot wave through a Mississippi apartment house where twenty-five hurricane partygoers decided to wait out the storm. The building imploded on impact. One person and the concrete foundation survived.[45]

"Today humans are playing too large a role in natural disasters to call them natural," said a 1984 report from the Swedish Red Cross. "People are changing their environment to make it more *prone* to some disasters, and are behaving so as to make themselves more *vulnerable* to those hazards." As the earth experiences its most dramatic climatic shift since the time of Julius Caesar, the forecast is more of the same: global warming and rising oceans will make major storms more frequent, their impact more intense.[46]

The toe of Louisiana will vanish according to that dismal forecast. Already an acre of delta marsh disappears every fifteen minutes. Acreage the size of Rhode Island subsides every fifty years. River mud compacts, sinking grasslands. Ponds become estuaries. Beach migrates and so does the river that has carved in its time five paths to the ocean. There is no way to hold a mudscape in motion. Engineers experiment nonetheless—holding for the sake of 2.1 million people on 3.3 million threatened acres, for the nation's largest fin and shell fishery, for nine ports, for a $4 billion a year tourist industry, for 70 percent of the winged commuters on the Great Mississippi Flyway, for 15 percent of America's oil and 20 percent of its natural gas.[47]

The battle to hold the alluvial delta now consumes more tax and engineering resources than it originally took to reclaim it. Therein lies the tragedy of safety innovations that promote unsafe behavior, of unknowable hazards and unintended effects. In the wake of Betsy we see that levees block water and silt needed to replenish lowlands, that dredging loosens the land by killing freshwater plants. We understand that the cost of maintaining a Mister Go ($12 million a year; another $3 million or more per mile for marsh restoration) can be much greater than savings from faster shipping. We know that a moving shore is not much of a problem until humanity tries to stop it, that concrete solutions to coastal erosion can steepen a beach by deflecting its sand supply. We realize, or we should, that there is no turning back from all that we've made of the marsh without risking economic disaster. Paradoxically, tragically, we also know that to build as before would be to risk a worse catastrophe.[48]

"It's ironic," says Robert Brown of New Orleans, formerly a Corps publicist, now a homeowner with a cracking driveway. "The system which brings prosperity and security to humans is literally costing them the earth beneath their feet."[49] Too young to remember Hurricane Betsy, he's learned its lesson nevertheless.

Response to Environmental Change

When human societies alter their environment they generally must adjust to the changes wrought. Ian Douglas speaks of the city as a system of integrated energy flows in which one change prompts additional changes.[1] But he recognizes the city as more than a physical system. It is also a social creation. By acknowledging both physical and human dimensions, he recognizes that policy and its implementation contribute to environmental change, which in turn leads to further adjustments in policy and the environment. If we consider the larger urban industrial corridor stretching from Baton Rouge to New Orleans as a sociophysical environment, one can begin to assess the human response to obvious environmental change—particularly with respect to industrial pollution.

Called, depending on one's point of view, the "chemical corridor" or "cancer alley," the lower Mississippi River from just above Baton Rouge to near the river mouth has become a major petrochemical processing region since 1940. Intensive manufacturing activity has produced sizable quantities of industrial waste and associated unpleasantries and hazards. Smoke, odors, hazy skies, foul-tasting or contaminated drinking-water supplies, declining property values, hazardous-waste sites, and health complaints all are byproducts of an industrial economy. Initially recruited and fully embraced by Louisiana's political leaders, the petrochemical industry has developed critics in the past few decades. The criticism stems from several factors. Public toleration for pollution has waned significantly since the 1960s as federal programs have stepped up monitoring and regulation of industrial wastes. With declining oil prices over the past two decades and reductions in the manufacturing workforce, petrochemical plants are no longer viewed as nuisances whose economic benefits far outweigh their detriments. With the undesirable byproducts now largely exposed and the belief in economic salvation *abandoned, society is adjusting to the current role of the petrochemical industry. This section deals with those adjustments.*

The chapter by Craig Colten traces a series of crises that shifted public attention from agriculture-derived pollution to industrial sources and the manner in which government responded to water-supply problems in New Orleans. Industrial growth in Baton Rouge also produced a series of pollution-related crises. Raymond Burby asks where local government was during the expansion of the petrochemical industry there and considers how planners cooperated with industry to the detriment of the environment and neighboring communities. Barbara Allen examines how community groups in the vicinity of massive industrial pollution sources come to understand the health risks and how this affects their efforts to secure a higher quality of life. Popular understanding often stands in contrast to professional assessments of risk and this creates an obvious tension between the public and policymakers. Finally, Hank Bart discusses the impact of industrialization on fish diversity underscoring the fact that environmental change impacts both natural and human systems.

Too Much of a Good Thing

Industrial Pollution in the Lower Mississippi River

NEW ORLEANS draws its drinking water from the Mississippi River. Throughout most of this century government and industry officials viewed the Mississippi's gargantuan diluting capacity as adequate to protect urban water consumers. The average flow of some 600,000 cubic feet per second that coursed by Baton Rouge and New Orleans had, in the minds of the authorities, an almost limitless capacity to assimilate discharges to the point that they were harmless. Not only was the river a bottomless "sink," the state touted its huge quantities of fresh process water to prospective industries that built plants along the waterway in the postwar period. Since 1950 the principal state pollution-control authority granted discharge permits to manufacturers with impunity. In the state agency's view, industry's veritable dribble of effluent could hardly affect the mighty Mississippi. While contesting pollution elsewhere in the state, Louisiana authorities took virtually no enforcement actions along the lower Mississippi until the 1960s. Starting then and continuing until the mid-1970s, a series of nationally significant controversies over industrial wastes convinced many that the river's ability to accommodate industrial pollutants had been exceeded.[1]

This chapter will examine the pollution-control efforts in Louisiana as reflections of several fundamental shifts in political influence and technical capabilities. Early pollution cases, decided by local juries and judges, sided with agricultural or fishing interests over big business. As industry, particularly oil production, rose in importance, appeals courts reduced fines against polluters or overturned the cases altogether. Beginning in the 1940s the Stream Control Commission directed equivalent attention to agriculture-related pollution—including timber processing—and oilfield wastes. Not until the 1960s did state authorities begin to challenge the petrochemical industries, and then only tentatively. It took federal involvement to identify out-of-state petrochemical pollution sources and thereby target those outlets. Additionally, the shifting emphasis in pollution control was not just from one industrial sector to another but had a decidedly geographic dimension. The change from protecting farmers and fishermen to guarding the state's major municipal water supply—namely, New Orleans—embodied a rural to urban reorientation and also an upstream to a downstream focus. Finally, techniques to detect small quantities of chemical contaminants, introduced by federal investigators, sparked greater attention to finished water supplies than to raw river water. Through a review of early state laws and nuisance suits, an analysis of state agency enforcement actions, and an evaluation of outside influences, this chapter will provide a historical and geographic assessment of pollution control on the lower Mississippi River and its relationship to New Orleans' drinking water.

HISTORICAL CONTEXT

Several fundamental currents flowed through pollution-control efforts in Louisiana and the lower Mississippi in the twentieth century. First, post-1945 industrial development in the South accelerated a process started during World War II and transformed the environmental quality of the region. Southern states embraced policies that encouraged manufacturers to build plants, and Louisiana was no exception. Like its neighbors Louisiana crafted favorable tax policies for new manufacturers. While sugar and paper mills and other industry based on local agricultural products or natural resources were scattered throughout the state and were augmented by a cluster of manufacturing in New Orleans, no major industrial complex existed before about 1950. As of 1951 the U.S. Public Health Service reported fifty-eight plants along the lower Mississippi, but only six produced petrochemical products. More than forty plants along the river worked with food or timber products. This ratio changed over the next decade as petrochemical companies acquired land on the riverbanks and built massive refining operations—statewide the number of petroleum and chemical plants rose from 172 to 255 between 1947 and 1967.[2] The concentration of plants along the Mississippi was most pronounced. By 1971 the U.S. Environmental Protection Agency reported sixty "major" petrochemical plants on the lower river—a tenfold increase from 1951.[3]

Second, as business historian James Cobb has noted, southern states did little to inhibit releases of untreated effluents during the industrial boom years. This policy created an increasingly obvious pollution problem. Rooted in regional economic development efforts, state environmental policies accommodated the new and welcomed payrolls. At a time of no specific federal pollution-control standards, state policies characterized a typical "race to the bottom" control approach. In order to remain competitive with their neighbors the states refused to erect pollution-control barriers around their borders.[4]

Third, the shift from agricultural and timber-processing operations to petrochemical plants also changed the fundamental nature of contaminants in the waterways. Biological wastes such as effluent from sugar-cane grinding operations and pulp and sulfite-laden wastes from paper mills placed a high biochemical oxygen demand on streams, but the Mississippi could accommodate it. The petrochemical plants were a different matter. Complex organic chemicals, brine (salty water), and oily wastes presented different problems, both for aquatic life and municipal water supplies. Nationally, public health officials had been expressing concern with phenolic and oily wastes since the 1920s. Discharges containing these contents produced foul-tasting drinking water, especially when treated with chlorine, or obvious and objectionable oil slicks. In the lower Mississippi, the huge diluting capacity of the river permitted the release of limited quantities of phenolic wastes, while plants with oily wastes could employ mechanical devices that separated the oil from the other effluents.[5]

In the early 1950s industry and public health agency research sought to define the toxic level of various contaminants in waterways.[6] Although many of the new wastes contained highly toxic ingredients, state agencies continued to permit releases, even upstream from New Orleans's water intakes. Concern focused initially on wildlife that was susceptible to smaller concentrations of toxic wastes but gradually expanded to include human health in the state's largest city. Through the use of more systematic, basin-wide water-quality sampling and new analytical methods, officials focused more on minute quantities of harmful organic chemicals found in finished water supplies. Ultimately, discussions about the effect of these contaminants played a considerable role in the passage of federal water-quality legislation.

EARLY POLLUTION CONFLICTS IN LOUISIANA

Louisiana case law indicates that early twentieth-century local courts and juries sympathized with farmers or other rural landowners who suffered damage from pollution, although appellate judges typically rejected claims for large damage payments. When oil and brine discharges from the several companies drilling in the Jennings oilfield in southwest Louisiana damaged property of William McFarlain in 1904–5, the local court awarded the plaintiff $1,000 in damages. On appeal, McFarlain requested payments totaling $20,0000, but the court affirmed the original verdict and refused the larger award—arguing the entire farm was not worth $5,000. Albert Long of Winfield sued Louisiana Creosoting in 1915, and the local court granted him a modest sum of $200 (of the $2,000 requested). When the plaintiff appealed the initial ruling, the state supreme court refused to raise the damage award. Similarly, in 1937, after a Sabine Parish court ordered the Pelican Natural Gas Company to pay a neighbor for damages to his stock pond and timber, the appeals court affirmed the ruling but reduced the amount from $3,400 to $1,000. Again in 1951, downstream property owners charged a paper company with damages to their land and livestock resulting from industrial discharges. The jury awarded one landowner over $19,000 in damages, but the appeals court reduced the award to $10,000.[7]

Statewide, the local courts sought to compensate those who suffered from industrial pollution while refusing to impose injunctions on manufacturing activity. The higher courts agreed that damages had occurred, but either approved low damage payments or reduced large awards. One judge reasoned "if industry is to be permitted a reasonable chance to develop for the benefit of the whole community, some inconvenience must be endured by its inhabitants."[8] This position granted industry limited rights to pollute, while not completely discarding traditional riparian water rights. The precedent-setting pollution cases involved incidents on smaller waterways where the impacts of early twentieth-century industry were most obvious and deleterious. No significant industrial wastes cases involved riparian landowners or cities along the lower Mississippi.

Louisiana initiated its statutory challenge to pollution with an act to protect agricultural interests while allowing oil and gas producers seasonal use of surface waters. The Louisiana legislature passed an act in 1910 that specifically protected farmers using irrigation water by forbidding discharges of oil, saltwater, or "other noxious or poisonous gases or substances that would render said water unfit for irrigation purposes or would destroy the fish in said streams."[9] The prohibition was valid only from March 1 to September 1 and permitted discharges during the fall and winter. The act stood as a compromise—protecting irrigation waters when farmers drew on them, but allowing oil producers to use waterways the balance of the year. It also focused attention on the southwestern portion of the state where rice cultivation used irrigation water. A few years later (1924), the legislature expanded pollution control and passed an act, without seasonal restrictions, prohibiting pollution by saltwater or oil. This bill made it a misdemeanor to kill fish in the state's streams throughout the year but made no mention of agriculture.[10] By placing enforcement authority with the Department of Conservation, the general assembly emphasized the wildlife protection dimension of pollution control.

The state took enforcement action in 1912, filing suit against an oilfield operation that had released brines and damaged rice crops. The oil-producing defendant claimed that the state statute (Act 183) was unconstitutional due to the fact that it discriminated against a certain type of business. The state supreme court, however, upheld the law, claiming that the law protected the general public and did not discriminate against any one industry.[11]

On the eve of its industrial buildup (1940), Louisiana created its Stream Control Commission (SCC). Modeled after similar organizations in other states, the commission comprised state officials from several agencies such as Conservation, the attorney general's office, and Public Health. Of fundamental importance, it received the power to restrict pollution of state waters. Specifically, the legislature delegated the authority to establish rules and regulations relating to water pollution that was destructive to aquatic life or injurious to public health. During its first few years of existence, the commission claimed its chief accomplishments were solving internal organizational problems and establishing a cooperative, rather than hostile, relationship among sportsmen, industrialists, and agriculturists.[12]

As was typical at the time, industry sought to play a leading role in pollution control—not necessarily to prevent it, but to help define pollution, to monitor water quality, and to recommend proper control procedures.[13] In Louisiana the big three industrial-pollution sources were petroleum, pulp and paper, and sugar. Both the Louisiana Petroleum Refiners Waste Control Council and the National Council for Stream Improvement (a pulp and paper industry organization) sponsored research on the effects of their respective wastes and how to control them.[14] The petroleum refiners' pollution-control organization actually sponsored biologists working for the state. These scientists carried out tests to determine the toxicity of industrial wastes. Such findings helped establish dilution ratios for receiving waters.[15] In addition, the Pollution Abatement Committee of the American Sugar Cane League worked with the state to minimize the impacts of sugar-house wastes.[16] Most pollution-control efforts concentrated in smaller watersheds and minor streams. Neither the state nor industry groups paid much attention to the Mississippi, other than conceding that it could handle most wastes dumped into it. Indeed, several sugar factories near Plaquemine promised to divert their effluent from small bayous to the Mississippi where it "would appear in inappreciable concentration."[17] Concern with industrial effluent commonly focused on a single waste stream. The state cautioned oil producers near Baton Rouge that their brine discharges could threaten New Orleans's water supply if they exceeded 10,000 barrels per day.[18] In 1947 Baton Rouge refineries asserted that their releases to the river did not threaten fish life, and that they separated oily waste and kept it from the waterway.[19] In subsequent years problems with oily waste emerged, but the state sought a cooperative rather than legal solution. In particular, one refinery's waste accumulated in a ditch leading to the Mississippi. When the river level fell rapidly, "oil accumulation of several months is discharged in a matter of a few hours." State officials worked with Standard Oil to achieve a solution that kept the oily wastes from the river. Closer to New Orleans, Shell Oil's Norco plant diverted its effluent, first to a pit and later via a canal to Lake Pontchartrain, with the state's approval.[20] This represented one of the few instances when a refiner excluded *all* its wastes from the Mississippi; it stems from Shell's proximity to New Orleans (about twenty highway miles upriver).

Placed in the state conservation agency, the SCC had a priority to preserve aquatic life. During its early years, the commission balanced its attention among oilfield, pulp and paper, and sugar-cane processing pollution sources that threatened the state fishing industry. The commission's first set of rules (1941) focused attention on oilfield wastes and prohibited the release of brines in concentrations that would affect the receiving water's palatability or ability to support aquatic life or livestock. It also required the separation of oil from brines before their release. This prompted a challenge from the oil industry. In a 1948 suit the Texas Company claimed that the SCC denied oil producers due process and treated them unfairly by restricting the industry's discharges in coastal areas with brackish waters. A federal court found that the commission had a legitimate authority to protect state waters and that it acted fairly in executing its duties. In its ruling, the federal court confirmed the state's authority to protect its vital fisheries from industrial effluent.[21]

In addition to industrial sources, the SCC sought to identify naturally occurring threats to aquatic life and to foster the belief that industrial sources were not the source of all fish kills. The commission claimed that well-meaning citizens attributed fish kills to industry when saltwater intrusion, temperature-related oxygen depletion, or other factors were responsible. Nonetheless, taste and odor complaints stemming from releases of oil into the Mississippi during the 1940s attracted the attention of the Louisiana Board of Health, if not the SCC.[22]

When the U.S. Public Health Service (USPHS) published its 1951 report on water pollution in the lower Mississippi, it concluded that "stream pollution is not serious except in a few local areas where industrial wastes cause problems." Most notably, it observed that "industrial wastes from Baton Rouge may affect the New Orleans water supply if the concentration of wastes becomes high enough."[23] Nonetheless, it noted that the volume of water was adequate to dilute existing discharges. The health service based its conclusion more on water-quality data at New Orleans than knowledge of actual industrial discharges. A dissertation on water pollution prepared at Tulane University two years later only considered bacterial standards and concluded that water taken from the river at New Orleans was safe to drink. This study's findings reflected the traditional emphasis on bacterial and biological contamination and the water-quality data available at the time. The USPHS report also pointed out that the release of untreated sewage from New Orleans prevented the consumption of shellfish growing downriver from the city. While investigators saw industry as a pollution source, they typically focused attention either on smaller streams or municipal sewage in the Mississippi River. The fact that refineries close to New Orleans diverted their effluent into Lake Pontchartrain to prevent foul-tasting water downstream encouraged this perspective.[24]

The state's inattention to industrial pollution sources along the Mississippi was emphasized by information passed along to federal officials in the late 1950s. When the USPHS compiled an inventory of industrial and municipal waste-treatment facilities on the Mississippi, Louisiana reported only five companies and many communities, large and small (1953 data). Each manufacturing firm reported discharging its wastes to a ditch or canal rather than the main stem of the river.[25] State records, therefore, indicated only a minor threat to the river. The U.S. Geological Survey (USGS) apparently saw industrial pollution as a greater hazard, reporting in 1956 that it was an "ever present danger" at New Orleans. It also emphasized the saltwater intrusion problem as a critical issue for New Orleans's water supply. Chemical analyses conducted regularly by the USGS included a variety of inorganic chemical constituents, but did not consider organic chemicals.[26] Federal concern with New Orleans's water supply appeared to be greater than state and local attention during the mid-1950s.

PERMITTING POLLUTION

In 1951 the SCC established requirements for industries discharging wastes into state waters. It mandated that industries apply for permits to release wastes into streams and created a process whereby the commission would approve discharges.[27] Thus, the many industries lining the Mississippi had to submit engineering evaluations of their planned releases for the state's consideration and approval. It was not until late in the 1950s that the SCC began keeping records of its hearings on discharge applications. Although inconsistent in terms of quantities reported, the hearings illustrated the state's willingness to allow industry to use the river for transporting and diluting effluent.

The SCC received applications for discharges to the Mississippi River at least thirty-seven times between 1958 and 1966. These applications included substantial discharges from new plants, effluent from minor process changes or plant expansions, and discussions of accidental releases. Companies reported the discharge quantities in a variety of measures: parts per million, gallons per minute/hour/day, and pounds per hour/day. Quantities ranged from a few pounds to over 200,000 pounds per day (Kaiser total solid discharges). Chemicals being released into the river included a variety of salts (including cyanide), organic chemicals (oil, kerosene, and acetone), metals (zinc and copper), inorganic materials (calcium), and total solids. Additionally, several companies requested permits to turn their cooling water into the river. The incompatible measures reported to the SCC indicated no effort to assess the cumulative effect of the many industrial discharges. The SCC considered the impact of each release separately within the total Mississippi River flow. Other than salt, the commission gave no apparent consideration to the presence of the same chemicals in downstream water supplies. Of the thirty-seven applications, the commission rejected only *one;* two applications represented second attempts. The one rejection stemmed from the commission having insufficient details about salt and phenol discharges from a plant just below New Orleans, and it requested more information. The commission was not a rubber-stamp operation, but it did little to inhibit discharges by manufacturers.[28]

In addition to permits, the SCC could restrict pollution by issuing abatement orders.[29] When the SCC tallied its abatement orders issued during 1950s, the object of the state's enforcement attention became obvious. The commission reported 163 abatement orders between 1950 and 1959. Only one incident represented action against an oil refinery, while the highest total belonged to oilfield producers. Sugar mills received the second highest number of abatement orders. This reflected strong public opposition to the nuisance created by the cane processors.[30] Overall, the commission focused much greater attention on oilfields and agricultural-products processors during the 1950s.

Public complaints about the offensive taste and odors in Mississippi River water nonetheless prompted an SCC analysis of that water in 1957. The investigation reflected increasing concern with the stream's inability to handle unrestricted quantities of industrial effluent. The commission's biennial report attributed some of the problem to naturally occurring causes but conceded that refineries and organic-chemical manufacturers produced objectionable wastes. The survey included chemical analysis and biological sampling.[31] In addition, the commission called on industries to submit weekly reports that tabulated toxic and taste-causing discharges.[32] The program prompted several industries to make adjustments in their waste-treatment systems. At least twenty-two manufacturers along the lower Mississippi voluntarily participated, including eighteen chemical or petroleum refineries.[33]

This did not solve the problem, however. In early 1960 an industrial spill into the river sparked a public outcry. Phenols in the river severely fouled public drinking-water supplies below Baton Rouge. This stimulated huge bottled-water sales and forced producers to discard cases of soft drinks produced with river water. The state responded by creating an emergency warning system to alert water-supply services of impending pollution. The plan required each plant to report any sizable spill to state authorities, who in turn would alert downstream water-supply facilities to close their intakes. Several spills in the ensuing months tested the system, and officials declared it a success.[34] Still, pervasive taste and odor problems convinced the SCC that the volume of industrial waste had surpassed the Mississippi's capacity to handle unlimited and untreated discharges. By commencing an ongoing monitoring system they were able to detect adverse changes and relate sources to taste problems and fish kills. Nonetheless, the basic strategy was to allow discharges and shut down water supplies when unusually large spills entered the river.

TOO MUCH OF A GOOD THING

After permitting industrial discharges with virtual impunity for nearly two decades, the SCC encountered three nationally significant problems that forced a reevaluation of industrial pollution. A series of massive fish kills during the early 1960s, followed by frequent taste problems in the late 1960s, culminated with a cancer scare in the early 1970s that shifted attention from aquatic to human life and also placed greater emphasis on industrial wastes as the most pressing problem.

In the winter of 1963–64 a massive fish kill in the lower Mississippi River riveted public attention on pollution caused by agricultural chemicals. Occurring shortly after the publication of Rachel Carson's *Silent Spring* and its indictment of chemical pesticides, this event attracted national attention. Louisiana had faced similar incidents in previous years when sportsmen complained of fish kills on small streams in the cane-growing region during the fall grinding season. The state sought to control the problem by restricting the times when farmers could apply pesticides. By banning the application of pesticides during the last forty-five days of the growing season, authorities hoped to minimize the residual endrin that would be released by grinding operations in the fall.[35]

The 1963–64 event garnered vastly more public attention than previous events, while the earlier kills prompted an indignant but local reaction.[36] Events in 1961 and 1962 were minor in terms of the number of fish affected (about 250,000 each year). The estimated number of dead fish in 1963–64 totaled about 5 million and it occurred after the pesticide application and cane-grinding seasons. Given the timing of this event and a public already attuned to pesticide pollution, it became a national event. The *New York Times* ran frequent stories throughout the spring of 1964, congressional hearings chaired by pesticide foe Senator Abraham Ribicoff, and regional hearings conducted by the USPHS gave the event additional visibility.[37] The initial assumption was that agricultural chemicals, including endrin, drained from farmland and wiped out aquatic life. Louisiana authorities, unable to isolate the cause and suspecting a source beyond the state's boundaries, called on the USPHS for assistance.[38]

In response to Louisiana's appeal, the USPHS initiated a search for the source throughout the reach of its monitoring territory, which included sampling stations as far north as Minneapolis. Its investigators promptly ruled out low river stage (minimal dilution), deficient oxygen supply, extreme temperatures, or viruses. Tissue samples of dead fish analyzed by health service scientists consistently revealed the presence of endrin—an insecticide used in the Louisiana sugar-cane fields and elsewhere in the Mississippi valley. Despite this finding, the USPHS's preliminary report echoed Louisiana's uncertainty about the source.[39]

Continuing its search, the USPHS reexamined water samples collected during previous years as part of its water-quality surveillance system on the Mississippi. The results showed spikes of endrin in the New Orleans water supply during the summers of 1960, 1961, and 1962. There were no comparable spikes upstream at Vicksburg. This suggested the previous events resulted from pesticide runoff from sugar-cane fields or discharges from cane-grinding operations. However, in the fall of 1963 and winter of 1964, a series of endrin spikes appeared to travel downstream from West Memphis to Vicksburg and New Orleans (their system reported no spikes further upstream) (see fig. 9.1). This pattern convinced investigators the source was the Memphis area, not the cane-growing region.[40]

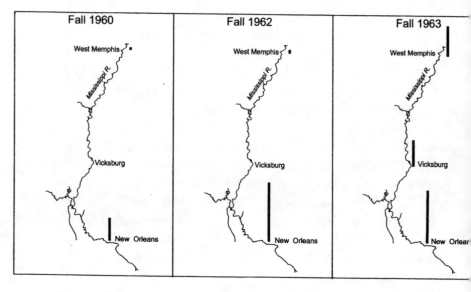

Figure 9.1. Relative concentration of endrin in the Mississippi River, 1960-1963.
(Source: Federal Water Pollution Control Administration, "Endrin Pollution," 1969.)

An elaborate series of studies coordinated by the USPHS ultimately pointed an accusing finger not at farmers or crop dusters but at the lone endrin producer in the lower Mississippi valley—Velsicol Chemical Company's Memphis facility. The USPHS quickly redirected the search from farmland to industrial wastes. Investigators dispatched to the Memphis Velsicol plant found high endrin concentrations in sludges lining the ditches and in streams where the company released its wastes. Furthermore, investigators suspected that seepage from a large municipal landfill where Velsicol sent sludges ultimately entered the Mississippi. In addition, a firm that recycled Velsicol's barrels allowed drainage from its rinsing operation to flow into sewers leading to the river.[41] What were obviously sloppy waste-management practices convinced the USPHS that they had identified the source.

Opponents of the pesticide industry longed to use the 1963–64 fish kill to curtail heavy agricultural chemical use. Nonetheless, the USPHS concluded that industrial wastes were the source and took steps to convince a confrontational Velsicol Chemical Company to clean up its operations. Velsicol, predictably, contested the accusations. Corporate spokesmen questioned how a highly toxic concentration of endrin could travel from near

Memphis to the lower river without killing more fish along the way. They also pointed out that Menhaden, a marine species and not river catfish, accounted for 96 percent of the kill, but the USPHS analysis of chemicals in fish flesh focused on the bottom-dwelling catfish.[42] Public Health Service investigators were not swayed by the corporate questions, and local authorities forced Velsicol to divert its wastes to rural land-disposal sites. While attaching blame to Velsicol's waste-disposal practices was disappointing to the foes of agricultural chemicals, the USPHS decision gave rise to a sense of relief in the chemical industry. Proponents of agricultural chemicals reasoned it would be easier to weather a single cleanup than to have Congress impose restrictions on the industry as a whole. Steps taken by Velsicol to divert its wastes to a remote land dump and decreasing use of endrin in subsequent years prevented further kills of a magnitude equivalent to the 1963–64 event.[43]

Three significant pollution-control trends emerged from this event: (1) industry became the object of primary concern, shifting attention from agricultural and municipal wastes; (2) the USPHS began to consider the Mississippi River as a large system where chemicals released into the waterway could cause damage substantial distances downstream (and not a set of unrelated local pollution cases); and (3) concern with human health emerged when the analyses found endrin in New Orleans water, although at levels considered insignificant to affect humans.[44]

During the next decade public officials continued to place human concerns higher on the priority list for pollution control—effectively this meant New Orleans's water supply. The Louisiana State Department of Health reported on the effectiveness of its early warning system but noted that the increasing load of industrial pollution defied municipalities' ability to remove all odor and taste contaminants.[45] By 1967 the state agency lamented the failure of its early warning system and called for "full-time pollution surveillance patrols in the Baton Rouge and New Orleans area."[46] "Chemical" and "oily" tastes attributed to river water and to fish caught below Baton Rouge had become such a chronic problem by 1967 that state officials no longer regarded them as evidence of discrete spills but as proof of overall water-quality deterioration. Furthermore, federal investigators in 1967 performed inconclusive analyses on New Orleans's water to determine if it contributed to cancer in laboratory animals.[47] This combination of concerns led Louisiana SCC's chairperson, Dr. Leslie Glasgow, to request assistance from the Federal Water Pollution Control Administration (FWPCA) in identifying the source(s).

The FWPCA (and later the U.S. Environmental Protection Agency [USEPA]) began an intensive analysis of pollution sources, focusing on industrial dischargers. The federal investigation extended from early 1969 to mid-1971 and sought to evaluate the pollution caused by industry downstream from St. Francisville, identify any water-quality violations, and if necessary help the state abate pollution. A preliminary report submitted to the U.S. Senate in the spring of 1971 concluded that the petrochemical industries were largely responsible for the taste and odor problems downstream from Baton Rouge. It also indicated that these discharges might pose a threat to downstream water consumers, to estuarine and marine life, and even to people who consumed oysters collected near the river.[48]

The USEPA released its full report in 1972, and it identified sixty major industries discharging wastes to the river (see fig. 9.2) and proclaimed that scientists had detected a total of forty-six organic chemicals in the three principal public water supplies (Carville, Jefferson Parish, and New Orleans). The investigation did not stop with the analysis of organic chemicals (the most likely source of undesirable tastes and odors) but analyzed the wastes for toxic metals and salts, as well as total solids that might contribute to odor and taste. The report took an unprecedented step by publishing the names of the contributing industries and their discharge volumes.[49]

The FWPCA/USEPA analyzed both industrial effluent and public water supplies. This research required use of relatively new analytical methods to identify the organic chemicals.[50] At least seventeen industries discharged more than ten pounds of phenols per day (a principal source of chemical tastes in water supplies). While the individual plant discharges were insufficient to cause taste or odor problems, the combination of over 8,400 pounds per day apparently exceeded the taste threshold—USPHS standards called for no more than 0.001 mg/l in water supplies. Permit evaluations carried out by the SCC did not consider the combined effect of industrial discharges. Therefore, the existing regulatory system made no provision for the cumulative effect of multiple discharges. By analyzing the public water supplies using new and powerful analytical techniques, the FWPCA/USEPA study shifted attention from individual plant discharges to the total downstream impact. While the 1964 USPHS study analyzed raw river water, the 1972 assessment made treated drinking water its prime concern.[51]

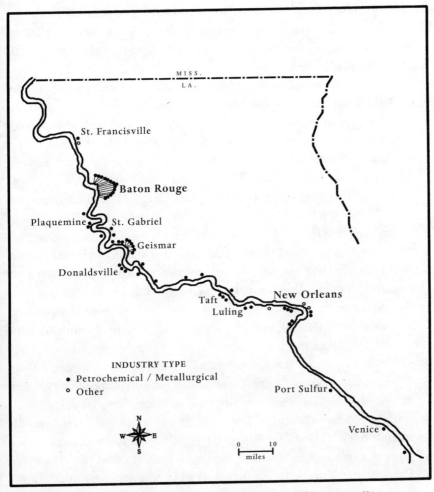

Figure 9.2. Industrial sources of pollution to the Mississippi River, 1972. *(Source: USEPA, "Industrial Pollution," 1972.)*

The report concluded that the industries jointly contributed "significant quantities of hazardous and/or undesirable pollutants to the Mississippi River in Louisiana and that these wastes were the principal cause of persistent oily-petrochemical tastes in downstream water supplies and in fish." Furthermore, it specifically pointed out that trace amounts of organic chemicals "are a potential threat to the health of the 1.5 million people who consume this water." Since existing waste-abatement measures were inadequate to prevent these discharges, the report advocated the removal and/or reduction of "hazardous discharges to the river." Additionally, it called on the municipal treatment plants to remove both organic and heavy metal contaminants.[52]

During the federal investigation, Louisiana requested industries to install additional waste-treatment facilities by 1972. The SCC reviewed all permits and pressured industry to treat their effluent for several of the specific organic chemicals identified by the USEPA, namely chloroform, benzene, and bischloroethyl ether.[53] This spurred a shift in disposal practices from stream releases to land disposal. Rural land areas that have since become Superfund sites began receiving the organic chemicals excluded from the river. The notorious Petro-Processors (PPI) sites north of Baton Rouge began accepting chemical wastes diverted from the river during this time. By 1970 spills from the PPI site had damaged neighboring farmland.[54] In addition chemical manufacturers turned to deep well injection systems amid concerns that land disposal would cause groundwater contamination.[55] Even at the outset of the diversion of wastes from water to land, public officials recognized the shift in disposal media but, instead of eliminating the hazard, merely relocated it.

In 1973 the Mississippi River system became the object of concern over drinking water quality. Ralph Nader and his scientific consultant, Robert Harris, of Johns Hopkins University, testified before Congress that drinking water supplies in Cincinnati, Ohio, and Evansville, Indiana, contained numerous organic chemicals. Nader suggested that high cancer rates in New Orleans were the result of a contaminated water supply. He added that New Orleans officials were reluctant to allow an associate to inspect the city's filtration system, implying they had something to hide.[56] In 1974 Robert Harris, then representing the Environmental Defense Fund, publicly charged New Orleans and the state of Louisiana with inadequately protecting the city's water supply. In a multipart article in *Consumer Reports*,[57] Harris asserted that New Orleans had an exceptionally high cancer rate and that this was due to an environmental source, which could be the drinking water. A fundamental objective of the article was to spur the use of improved treatment of drinking water, particularly activated charcoal systems. The articles were not a direct attack on New Orleans. Harris conceded that New Orleans's water was not "necessarily worse than that of most other cities." He also allowed that "some other water supplies may be in even greater need of activated charcoal treatment."[58] While a tempered criticism of New Orleans's water supply, it stirred resentment there and stimulated a lively national debate. The atypical outlet for a scientific discussion enshrouded this episode in professional controversy but certainly captured public attention. Local officials and the USEPA quickly responded that the local water supply was safe.[59] Nonetheless, the debate stimulated the sale of bottled water and one report even charged that City Hall relied on this source while proclaiming tap water safe.[60]

Appearing about a week before a congressional vote on the Safe Drinking Water Act (1974), a second Harris report charged that the cancer mortality rate for white males consuming Mississippi River water was 15 percent higher than for those consuming other water supplies. Harris claimed this suggested a link between the river water and cancer, although he admitted it was not conclusive. A New Orleans paper accepted Harris's speculations and proclaimed that proper treatment would end fifty cancer deaths a year.[61] The ensuing public debate contributed to a strong House vote on the Safe Drinking Water Act (1974), which offset a threatened presidential veto (this act allowed the USEPA to establish limits for chemical contaminants in public water supplies or required treatment methods to remove specific chemicals). In addition, the Harris report stirred up local officials and the scientific community, both of whom questioned his conclusions.

Public health scientists challenged the assumption that the chemicals in the water supply were carcinogenic, since many had not been evaluated. Also, critics called for detailed studies of cancer patients to determine their actual exposure to river water and other environmental factors.[62] About half of the parishes with high cancer rates included in the Harris analysis (four of nine) did not use Mississippi River water. Critics suggested this was an obvious flaw. Others found fault with the conclusion that New Orleans needed to substitute an activated charcoal filter for the sand filtration system used to treat its water supply. Tests of raw water and finished water revealed that several organic chemicals passed through charcoal filters, one even increased in concentration.[63] The National Institutes of Health evaluated the Harris report and stated that "one can (and should) write about potential carcinogens in drinking water in a fashion that is coolly analytical rather than alarmist or passionate." Chief among their points of criticism was the study's focus on males, while excluding females who had a low cancer rate in Louisiana.[64] In a detailed analysis of the Harris report, scientists with the National Cancer Institute also pointed out that some of the same chemicals found in the river also occurred in Baton Rouge's groundwater. This, they argued, weakened the comparison of populations' drinking river water with groundwater supplies.[65]

Harris responded to his critics with an article in *Science* where he re-asserted his previous claims.[66] James Coerver of the Louisiana Department of Heath replied, arguing that Harris had been grandstanding, had refused to allow the state to review his report, and noted that state records showed high cancer rates in parishes that did not use Mississippi River water. Furthermore, a comparison of a functioning water-treatment system that filtered Mississippi River water through the prescribed activated charcoal system with New Orleans water showed almost identical concentrations of chloroform.[67] Nonetheless, Congress had already passed the Safe Drinking Water Act and New Orleans water had a sullied reputation.

This debate did not stimulate modifications in the New Orleans water-treatment process, but it caused a national stir and fostered legislation seeking to protect public water supplies. It also concluded the gradual shift in pollution-control emphasis from aquatic life to human well-being on the lower Mississippi. The New Orleans water controversy focused on chemicals in the municipal water supply, not on raw river water. There was no discussion at this phase of threats to aquatic life or of nonindustrial pollution sources. By drawing attention to municipal water supplies, the debate centered on the cumulative effect of multiple pollution sources and the inability of treatment works to remove these unwanted chemicals before their delivery to households. One additional shift found in this event reflected the attachment of blame to government bodies for failure to protect the public, rather than industrial concerns for introducing chemicals to the environment. This completed the cycle of concern from rural to urban and from wildlife to human.

CONCLUSIONS

New Orleans water was hardly a concern in early twentieth-century state efforts to protect waterways—although the city was certainly interested in a safe supply. The Louisiana legislature provided specific protection for farmers in 1910. The 1910 statute and subsequent common-law rulings favored agricultural uses of water over the newcomer—oil production. Appellate courts, although siding with farmers and other rural riparian water users, refused to grant large awards for damages caused by oil or other industrial discharges to the state's waterways. Before 1940 there were no precedent-setting cases involving the Mississippi River due to its huge capacity and the relatively small amounts of effluent released to it.

When Louisiana formed its Stream Control Commission in 1940, its initial mandate was to protect aquatic life, mostly in small streams and wetlands outside the urbanized portions of the state. The commission's early efforts focused on inducing cane mills, paper plants, and oil producers to minimize damage to waterways. The obvious thrust was to effect a cooperative program where the industries would determine what constituted pollution and develop means to prevent it. Although there was some concern with brine and industrial effluent discharges to the Mississippi below Baton Rouge, New Orleans's water supply remained in the background.

When the SCC began issuing permits for industrial releases to the Mississippi, it only considered the impact of each discrete discharge, not the cumulative effect of them all. Many small discharges, allowed under this policy, ultimately contributed to regular taste and odor problems in the 1960s. When foul-tasting water was combined with the disastrous fish kill of 1963–64, state officials recognized that the consequences of unimpeded releases extended beyond aquatic life. By the early 1970s, when the state called for improved industrial treatment and the diversion of troublesome waste from the river to land disposal sites, it indirectly offered protection for New Orleans's water supply. But environmental groups pointed to New Orleans as a water-supply disaster by the mid-1970s. Offered in a highly contentious manner in a politicized forum, these charges served as a catalyst for revamping federal policies on public drinking water, although they had little immediate impact on treatment locally.

From the great fish kill in the mid-1960s to the cancer scare in the mid-1970s, the lower Mississippi was a focal point for public debate over water quality. The rapid shift in concern from aquatic life to human health was embodied in the high-profile incidents on the lower Mississippi and a large urban population consuming the degraded water. After all, if the Mississippi River could not handle the industrial discharges released into it, all other streams were in obvious peril.

Raymond J. Burby

Baton Rouge

The Making (and Breaking) of a Petrochemical Paradise

S OME THINGS never change. In 1891:

> Irritated by the foul stenches that wafted through their northeast Brook-
> lyn neighborhood, members of the Fifteenth Ward Smelling Committee
> embarked on a boat trip up Newtown Creek in September . . . in search of
> the responsible parties. As their tug negotiated its way among cargo ships
> and manure scows, the odors were more pronounced until they reached a
> point across from the oil refineries where, according to the committee's
> report, "the stenches began asserting themselves with all the vigor of fully
> developed stenches." What the Smelling Committee quickly discovered
> was that an unusually heavy concentration of industrial activity—oil re-
> fining, chemical production, glue making, and fertilizer manufacturing—
> had transformed the area around Newtown Creek into an ecological
> wasteland.[1]

Oil tanks and refineries long ago left Brooklyn as the American petrochemical industry shifted toward the Gulf of Mexico and westward to California.
As the experience of Baton Rouge recounted in this chapter illustrates,
however, the problems the petrochemical industry creates for nearby residents have changed little over the intervening century. Numerous aggrieved
residents have recounted the day-in and day-out difficulty of living near

heavy industry in Baton Rouge.[2] In addition, systematic social science research shows that these burdens are disproportionately borne by black residents.[3] What has not been recounted before is how this situation came to
pass during a century in which national environmental regulations were put
in place to control pollution at its source and local planning and zoning regulations were developed to protect residential areas from industrial harms.

In this chapter I examine the industrialization of Baton Rouge, the state
capital, educational center, and most heavily industrialized city in Louisiana.
I look at the decisions that led to its transformation from a sleepy southern
town into one of the nation's leading petrochemical complexes, documenting the pollution and social distress that resulted. The story of twentieth-
century Baton Rouge is similar to those of nineteenth-century American
cities recounted by urban historians, such as the tale of the oil industry in
Brooklyn in the 1890s.[4] Like Brooklyn and other cities a century ago, in
Baton Rouge considerable environmental devastation has followed in industry's wake, environmental harms have fallen disproportionately on ethnic
minorities and the poor, and planners, government, and the legal system
have done little to right these wrongs.

THE BACKDROP: INDUSTRIAL POLLUTION IN URBAN AMERICA

Throughout the nineteenth century citizens and local governments waged
a war in the courts to curb the obnoxious practices of industrial establishments that fouled the water, air, and land, and, in many cases, made nearby
residential neighborhoods virtually unlivable. Their primary tool, the nuisance suit, however, had a number of flaws.[5] Frequently property owners
had difficulty proving which polluting industry, of those located nearby, had
caused them harm. To bring a public nuisance suit, local governments had
to be persuaded to take action, which required political skills and a degree
of influence poor and ethnic neighborhoods often lacked. Finally, as the nation industrialized, courts realized that economic growth could not occur
without some degree of pollution. Rather than shut down industrial establishments that were making life in nearby areas unbearable, the courts
instead ordered that compensation be paid for the harms they caused.
Determining adequate compensation, however, could prove difficult, and
the courts generally refused to pay for the social and psychological costs imposed by nuisances. Finally, the nuisance suit is reactive rather than proac-

tive. It is used only *after* a harm has occurred and does little to prevent obnoxious activities from locating in inappropriate areas.[6] In response to these deficiencies, urban reformers proposed a variety of approaches to industrial pollution that would, they hoped, prevent problems from arising. These approaches included remedies involving both land use arrangements and pollution control.

Isolation and Separation: The Planning Remedy

The first proactive approach many cities took was to ban obnoxious uses, such as slaughterhouses, rendering plants, and the like from the city.[7] Such ordinances can be found as early as the seventeenth century in the United States, but citizens were not the only ones who wanted obnoxious uses banned to the countryside.[8] Businesses as well often sought refuge from angry neighbors by moving to suburban locations.[9] In fact, industrial development in the suburbs, of which much has been made in the years since World War II, was well underway in the latter half of the nineteenth century.[10] But both the city ordinances and good intentions of industrialists typically failed to resolve the problem since, as cities grew, they placed households back into contact with the polluters. As a result, at the close of the nineteenth century, congestion of factories and residences remained a critical problem.

Because "neither city authorities nor businessmen did much to confine factories to industrial areas or to segregate the most offensive industries from residential communities," reformers looked to a land-use solution to solve the problem.[11] Early planning advocates, such as Benjamin Marsh and John Nolen, argued for decentralization of both factories and workers from highly congested and polluted central areas as an appropriate solution to the problem.[12] Following World War II the American Public Health Association's influential publication, *Planning the Neighborhood*, published in 1948, insisted that strict separation of housing from a variety of local hazards and nuisances was essential for public health.[13] The way to accomplish the needed segregation, according to planning texts of the same era, was to relocate heavy industry to planned industrial districts or parks, which, nevertheless, were easily accessible to the homes of industrial workers.[14]

Source Controls: The Engineering Remedy That Planners Embraced

In addition to separation, throughout the nineteenth century (and to this day), control of pollution at its source has been looked to as an important means to deal with environmental hazards from industry. In 1840 in England and beginning in the 1880s in the United States, local governments began adopting smoke-control ordinances and hiring municipal inspectors to see that industrial establishments complied. By 1962 a survey by the National Coal Association indicated that 140 of the 216 urban areas east of the Mississippi River had ordinances.[15]

After World War II city planners saw the potential to meld pollution-control requirements into zoning regulations. In a planning information report, the American Society of Planning Officials noted in 1951 that eleven adverse effects of industry could be controlled through industrial zoning: noise, smoke, odor, dust or dirt, noxious gases, glare and heat, fire, industrial wastes, transportation and traffic, aesthetics, and psychological stress. In fact, in the same report the society's writers suggested, "Many modern industrial plants would be an asset to many residential districts" if cities employed appropriate performance standards in regulating industrial land uses.[16] Urban economist Dorothy Muncy echoed this thought several years later in the *Harvard Business Review* when she observed, "If an industrial plant is designed to meet high performance standards, it should be allowed greater freedom of location in the community."[17] Thus, to the extent that separation proved to be unworkable, planners saw source control of pollution as a way to deal with heavy industry while still protecting the integrity of residential neighborhoods.

While planners were attempting to modernize zoning ordinances by incorporating both separation and source-control features, however, responsibility for regulating industrial pollution began to shift to the state and national levels. Political scientist David Welborn calls this an example of "conjoint federalism," since it involves sharing of power and responsibility across layers of the U.S. federal system. According to Welborn,

> Rising apprehension about environmental risks in the 1960s and 1970s and an increasingly sophisticated understanding of them provided a climate for national action, and the national government emerged as the dominant force in environmental regulation. Altogether, twenty-five legislative measures were enacted between 1961 and 1980 for the purpose of addressing problems of water and air pollution, solid and hazardous waste, noise, chemicals, land use, and the overall environmental impact of the national government's own activities.[18]

Thus, throughout the twentieth century a set of tools developed and were perfected that enabled planners, had they desired and garnered the political support to do so, to realize the aspirations for uncontaminated cities expressed by the planning reformers who came together at the first national planning conference in 1909. The extent to which these dreams were realized is examined in Baton Rouge, Louisiana, where industrialization began, coincidentally, in 1909.

Big Oil Comes to Baton Rouge

Wildcatters brought in the first oil well in Louisiana on September 21, 1901. John D. Rockefeller brought in the first oil refinery just eight years later when Standard Oil (now Exxon) announced that its agent, Colonel Frederick Weller, had purchased a 225-acre cotton field two miles north of Baton Rouge. There the company would break ground on April 13 for a plant to refine oil piped in from its midcontinent holdings 270 miles away in Oklahoma.[19]

Shut out of Texas by both ill feelings created by the shenanigans of a Standard Oil subsidiary and general antipathy toward the Standard Oil Trust, the company found a warm welcome in Baton Rouge. At the groundbreaking of new port facilities to accommodate the refinery, a company official remarked, "I appreciate the feeling here toward the Standard Oil Company as expressed by the remarks of Mayor Grouchy. The Standard Oil Company so often is referred to in a critical spirit, that I am glad to find it appreciated by the people somewhere."[20] The welcoming political climate was important, but it was not the primary reason for the choice of Baton Rouge. The site itself was ideal for a refinery. Located on high ground free of flooding, it fronted the Mississippi River and was bordered on the other three sides by the railroad lines of the Illinois Central, Frisco, and Louisiana Railroad and Navigation Company. Baton Rouge was at the head of navigation on the Mississippi for oceangoing vessels (a channel maintained by the U.S. Army Corps of Engineers connected it to the Gulf of Mexico 235 river miles downstream). This eased the import of crude from Mexico (and later Texas and Venezuela) and export of refinery products. The city also was accessible to the rapidly developing crude oilfields in northern Louisiana and Oklahoma.

Since 1909 the Baton Rouge petrochemical industry has centered on the original Standard Oil refinery.[21] Beginning with 700 workers, employment at the refinery grew to over 9,000 by the end of the 1940s (about a quarter of the parish workforce).[22] The area occupied by the plant increased over the years to encompass well over a thousand acres in two locations, the central refinery north of downtown and the Maryland tank farm a few miles to the northeast. Production grew steadily from a small plant running only 2,000 barrels of crude oil a day to produce illuminating oil, principally for foreign export, to its current run of over 450,000 barrels a day to produce over 700 petrochemical grades and products. As the oil industry shifted from the northeast to the Gulf Coast, by 1924 the Baton Rouge refinery had displaced Standard Oil's (Jersey Standard after breakup of the Standard Oil trust in 1912) flagship refinery in Bayonne, New Jersey.[23] Production at the Baton Rouge plant then increased steadily prior to World War II to 90,000 barrels a day in the 1930s and then increased again as the refinery became an important source of aviation fuel during the war. At the end of the war crude runs topped 137,000 barrels a day and then grew to 335,000 barrels in the 1960s, and to over 450,000 barrels a day by the 1980s.

THE EXPANDING PETROCHEMICAL COMPLEX

Expansion of the petrochemical industry in Baton Rouge came in three phases, each at a location further and further north of the city. These plants then typically expanded production as consumer demand for the products they produced mounted, and the city and state governments imposed few constraints on their operations.

The first phase, from 1909 to the early 1950s, centered on the Exxon refinery and included firms producing products needed by the refinery (e.g., sulfur, tetraethyl lead, chlorine, electricity) or by the war effort (e.g., aluminum). The dominant impetuses for growth during this period were the investment strategies of Exxon and the massive investment of the U.S. government in war-related industrial production. The second phase, from the early 1950s to 1990, occurred to the north near the historic black community of Scotlandville. During this phase, plants that use the various products of the Exxon refinery located in Baton Rouge in order to produce a vast array of products based on petrochemical feedstock (e.g., vinyl, plastic, rayon, foam rubber, and Styrofoam). The dominant impetus here was the synthetic organic chemical revolution and consumer demand for the vast array of resulting products. The third phase, which began during the 1960s and took place near the predominantly black suburban community of Alsen, includes large hazardous-waste treatment facilities. The impetus here has been federal regulations that require companies to treat their wastes and the exhaustion of on-site capacity for waste treatment at the various chemical plants to the south.

In addition to raw materials (and a market) provided by the Exxon refinery (by 1960 eight major manufacturing plants drew their raw materials through pipelines connected directly to it), petrochemical firms were attracted to Baton Rouge by a number of other business considerations. The Mississippi River provided an inexpensive transportation route, an abundance of process and cooling water (300 billion gallons a day flow by Baton Rouge), and a ready sink for industrial wastes (that same 300 billion gallons provided dilution of wastes). Excellent rail facilities supplemented river transport. In addition to an apparently limitless supply of petroleum and natural gas, raw materials such as sulfur and salt were readily available. The labor force was hard working, reliable, and cheap, though uneducated (which required investment in remedial reading and other skills development). Land for plants and ample additional room for expansion was readily available, first in the Exxon cluster area, and later in the vicinity of the Scotlandville cluster of firms. As additional plants located in the industrial complex, symbiotic relationships among plants developed, as did economies of scale as pipeline networks brought in crude oil and natural gas from newly discovered fields in south Louisiana and the Gulf of Mexico and interchanged products among complementary production plants. In the final stages of industrial development in Baton Rouge, specialized hazardous-waste disposal facilities, such as the large facility located near Alsen, just north of Scotlandville, also aided industrial location and expansion. Finally, the fair climate, made bearable with the advent of air conditioning, minimized downtime due to winter storms.[24]

Beyond business considerations, Baton Rouge provided a benign political climate that fostered rapid industrialization. Urban historian Andrew Hurley observed, "Historically, city governments in the United States have tended to make economic growth their first priority, whereas the specific strategies employed to attain growth have varied considerably according to time and place."[25] In fact, Robert Fisher notes that in almost all the cities of the Sunbelt a business elite controls local government decision-making, at least as it affects business operations and profitability.[26] Like its counterparts elsewhere, and particularly Houston, which it hoped to emulate, Baton Rouge accommodated industry in two ways. Arguably most important, it gave the petrochemical industry free reign to modify the environment as it saw fit. The industry was firmly established well before the first planning commission was appointed and zoning ordinance adopted in a fit of postwar progressivism. (Even forty-two years later, and after a series of environmental disasters, the consolidated city-parish's fourth master plan would report, "At a local level, the City of Baton Rouge Parish and East Baton Rouge Parish have few environmental ordinances or regulations.")[27]

In subsequent years, however, it also took numerous positive steps to encourage industry. Perhaps the most astounding came with adoption of a consolidated plan of government in 1947 (which became effective on January 1, 1949).[28] The plan divided the parish into three areas of governance. The urban area would have full city services and would be governed by the Baton Rouge City Council. The rural would not have city services and would be governed by both city and rural elected officials, sitting as the East Baton Rouge Parish Council. The third, termed "industrial," would also lack urban services (companies would provide roads, police and fire protection on their own), but property owners' tax liability would be capped at 4 mills per dollar of assessed value (i.e., a maximum property tax of $0.004 on each dollar of assessed value). The "provide your own services" provisions of the governmental plan were soon violated, however, as industries began to dump their wastes, solid *and* hazardous, in the city's Devil's Swamp landfill, located in north Baton Rouge.[29]

In addition to taking care to do absolutely nothing to harm or discourage industrialization, the city-parish was proactive in fostering development. To provide more land for industrial expansion it successfully lobbied Congress in the 1940s through 1960s for funds for a barge canal to serve the area north of the Exxon Group of industries. To reduce plant location and expansion costs, in the mid-1960s it formed an industrial development board and issued revenue bonds to finance the construction of plants for industries such as Allied Chemical. To ease the financial burdens of compliance with federal environmental regulations, in the 1970s it issued revenue bonds to finance pollution-abatement equipment. To allow industries to qualify for breaks under the 1980 Louisiana Enterprise Zone law, it established enterprise zones the law authorized. The state of Louisiana offered tax exemption for ten years in return for creating jobs. Exxon was one of the leading beneficiaries of the state program, garnering $93 million in tax exemptions during the 1980s.[30]

With industrial promotion came rapid population growth. Spurred by federal investment in Baton Rouge industries during World War II, the city more than doubled in population during the 1940s, when, along with Orlando, Florida, it achieved metropolitan status. Succeeding decades saw slower but nevertheless impressive population growth rates: 45 percent during the 1950s, 24 percent in the 1960s, and 28 percent in the 1970s. With the oil glut of the 1980s and general stagnation in "oil-patch" cities, however, the good times came to an end. In the good years, households poured into Baton Rouge from surrounding parishes. In the 1970s residents began to flee to adjacent parishes as they sought a higher quality environment in which to live.[31] By the end of the following decade, growth ground to a halt (4 percent for the entire decade of the 1980s).

THE ENVIRONMENTAL CONSEQUENCES

Burgeoning petrochemical production prior to 1950 brought numerous changes to the city. When Standard Oil arrived in 1909 Baton Rouge was a small town hosting the bureaucrats and legal functions of Louisiana's state government, which had moved back there from New Orleans after the Reconstruction in 1879, and an assortment of businesses serving the surrounding cotton plantations. The new industrial growth transformed the city.

Figure 10.1 is a bird's-eye view of the state capitol building, built by Governor Huey Long in 1932, and the Exxon refinery complex just to the north. The petrochemical complex, in turn, led to a sprawl of residential subdivisions east of the refinery as developers responded to the rapidly expanding market created by new industrial workers. The oddly named "Scenic Highway" buffered the new residential areas from the refinery. But, lacking both a planning commission and zoning ordinance prior to 1950, officials paid little attention to orderly growth and repeated most of the mistakes of industrialization in the Midwest and Northeast.

In the first comprehensive plan prepared for Baton Rouge, planning consultants Harland Bartholomew and Associates characterized land use in the city and parish as a "hodge-podge and chaotic pattern with residential, commercial, industrial, and public and semi-public uses all intermingled."[32]

Fig. 10.1. Louisiana State Capitol Building and Exxon Refinery to the north.

A 1958 report by the city-parish planning commission noted that while population had increased by about 40 percent since the Harland Bartholomew study a decade earlier, developed land use had increased by 82 percent.[33] This sprawling pattern continued toward the east of downtown and to the south as well, as more affluent residents distanced themselves from heavy industry. With residential growth on the natural levees of the Mississippi River to the north blocked by heavy industry, Baton Rouge expanded away from the river's natural levees and out into highly flood-prone back swamps. Ironically, many of these also were prone to pollution, since they lay at the outfalls of drainage works constructed to carry highly polluted stormwater from the downtown and Exxon industrial complex.

By the mid-1950s Baton Rouge residents began to notice smoggy conditions and complain of having to scrape particulate matter from their vehicles every morning. The city-parish president asked the local industries to pool their resources and monitor pollution. As might be expected, industry's response, after a year of monitoring air quality, was not to worry. The air quality data, according to an industry spokesperson, showed that air in residential areas was "relatively clean" and well within public health standards. In fact, he reported, while Baton Rouge certainly was not pollution-free, it was no worse, and better in some respects, than other cities with similar concentrations of heavy industry.[34] The air-pollution problem did not go away, however.

In spite of increased investment in pollution-control equipment,[35] by the mid-1960s the Baton Rouge Airport Commission complained that industrial pollution posed a hazard to flying and had caused the closing of Ryan Field on a number of occasions due to poor visibility caused by smoke and other industrial pollution. The commission suggested a system of industry self-policing might solve the problem.[36] The 1980s and imposition of federal air-quality regulations saw Baton Rouge under a federally imposed moratorium on new point-sources of air pollution (over 100 tons per year) due to its continuing status as a nonattainment area for ozone. The industry and governmental response, however, was to continue to study the problem by appointing an industry-government task force on air pollution in the late 1980s.

Dramatic reminders of public vulnerability have periodically visited Baton Rouge in the form of explosions and sudden releases of toxic chemicals. Small fires and other incidents at the refinery and chemical plants drew little public attention through the first five decades of the century. However, in 1965 an indication that something might be awfully wrong came to light in the wake of Hurricane Betsy, which rumbled through Baton Rouge in early September. Citizens woke up the day after the hurricane to learn, if they looked closely at the post-hurricane articles in the local paper, that a barge carrying 600 tons of chlorine was missing on the Mississippi River, which, as a precaution, had been closed to shipping twenty miles above and below the city. Interestingly, papers in other cities saw the issue more dramatically. Headlines in the *Winston-Salem Journal* screamed: "600 Ton of Chlorine Threatens Baton Rouge—Lost Gas Could Kill 40,000." Baton Rouge citizens became more concerned, however, when they learned that the federal government had shipped 116,000 gas masks to the city, none of which would fit small children. To deal with this crisis, the state civil defense director urged "very, very strongly" that anyone living within six miles of the site near downtown where the barge sank leave the area the day it was raised. All families within a twenty-five-mile radius were advised to "take a trip" for the day.[37]

Other startling indicators that something was wrong came periodically over the next two decades as tank-car derailments and fires at oil-storage facilities led to evacuations of surrounding neighborhoods. Possibly the crowning event of this type was an explosion at the Exxon refinery on Christmas Eve, 1989. The blast, which could be heard for miles, spewed asbestos over the surrounding residential area and damaged more than a hundred homes (Exxon paid damages to 150 homeowners).[38] In newspaper accounts following the accident, readers learned that Exxon had failed to contact or coordinate with local emergency response officials after the explosion. Equally disturbing, they found out that state emergency response personnel, who were hundreds of miles away responding to another accident, learned of it when they read the newspaper the next morning.[39] Concerned by these events, the Garden City Community Alliance, representing residents in the adjacent neighborhood, asked Exxon to purchase their homes. Exxon refused, but in subsequent years it quietly began buying the houses and clearing the area around the plant.[40]

In addition to explosions and widespread air pollution, both of which caught the attention of much of the urban population, the residents of Baton Rouge living in low-income, often black, neighborhoods had to cope with more localized forms of pollution that affected them in very real ways.[41]

Localized hazards such as odors, noise, dust, and other particulate fallout received little attention in local newspapers (and virtually no attention from local public officials) prior to the mid-1970s. At that point, however, increased political mobilization of black residents following the 1964 Voting Rights Act and civil rights movement led residents to be much more vocal in expressing their grievances.[42] Environmental justice soon became a rallying cry among activists fighting pollution in Baton Rouge.

The racial composition of census tracts in and surrounding the three industrial clusters (Exxon, Scotlandville, and Baker/Zachary/Alsen) is summarized in table 10.1. These data suggest why black residents began to feel victimized by industrial pollution. The workers who bought new homes adjacent to the Exxon Group of plants prior to 1970 were overwhelmingly white, but during the 1970s these neighborhoods increasingly came to be occupied by black households. That trend continued in the 1980s, so that by 1990 each of the neighborhoods adjacent to the Exxon Group of industries was predominantly black. In the case of the Scotlandville and Zachary/Baker/Alsen areas, industrial development did not get underway in earnest until after 1950. The neighborhoods most affected were then and continue to be predominantly black. The degree to which racial transition occurred because of the increasing frequency of environmental problems with the refinery and chemical plants, however, is not known.[43] What is known is that blacks mounted an intensive but not very successful campaign to right what they viewed as discriminatory industrial siting policy and discriminatory enforcement of pollution-control laws.[44]

During the 1980s Baton Rouge, with fewer than 400,000 inhabitants, ranked fifth in the nation, behind New Orleans, Houston, Los Angeles, and Chicago, in accidental releases of toxic chemicals. It even topped its mentor city, Houston, in one category of harm. Baton Rouge toxic incidents produced 171 injuries during the 1980s versus a total of 70 injuries in Houston.[45] Because of a federal law that forced companies to disclose their releases of toxicants to the atmosphere, by the end of the decade these grim statistics came to be widely known.[46] Residents learned that in a typical year, for example, Exxon's Baton Rouge chemical operation spewed 60,000 pounds of known and suspected carcinogens into the city's atmosphere or transferred them off-site. The top nine chemical producers released over 150,000 pounds known or suspected carcinogens per year. About 10 percent of the toxicants were released directly to the air, water, or dumped in pits at the production

Table 10.1. Racial Composition and Transition in Census Tracts Containing and Adjacent to Concentrations of Heavy Industry, Baton Rouge, Louisiana, 1960-1990

Census tract number	Percentage of population black			
	1960	*1970*	*1980*	*1990*
Exxon Group				
Tract 1	0.0	0.4	43.0	75.0
Tract 3	5.5	7.0	44.5	79.6
Tract 5	8.3	15.7	68.5	90.7
Tract 8	0.0	1.5	79.8	86.6
Tract 30.02	99.1	99.0	99.5	92.3
Scotlandville Group				
Tract 30.01[a]	99.1	99.1	99.2	99.8
Tract 31.01	98.7	96.3	94.7	99.8
Tract 33	37.0	17.9	96.0	98.7
Baker/Zachary/Alsen Group				
Tract 30.01	99.1	99.1	99.2	99.8

[a]This tract contains Southern University, a traditionally black institution of higher education.

site. The other 90 percent went to nearby waste-disposal sites located in or near predominantly black neighborhoods.[47]

Two grievances, both of which were related to the disposal of hazardous industrial wastes, drew national attention and a sustained campaign by environmental activists to redress harms. The first stemmed from the release of hazardous chemicals from Petro-Processors, a chemical waste-disposal firm that had poured petrochemical wastes into pits located a short distance from Devil's Swamp Landfill in the Scotlandville area. In the winter of 1969–70, one of the pits broke open and released tons of toxic wastes into the swamp. The accident came to light when a nearby rancher, David Haas Ewell Jr., found a number of his angus cattle dead and dying after they wandered into the area in search of food. It came to public attention in 1986 when *Science Journal* publicized the event and Ewell's decade-long quest to have the land, which had been in the family for more than a century, restored so that he could once again use it.[48] The case went to trial in 1975. As was typical in nuisance suits, the judge ordered compensation for the loss of 150,000 acres and the 150 cattle that had died ($25,000 and $5,000 for mental anguish), but he did not order cleanup of the property. Ewell had to wait another five years before the Louisiana attorney general filed suit under federal hazardous-waste laws to force the ten companies whose waste contributed to the mess to pay to clean it up. This process then took a number of additional years to get started.[49]

The second case involved Alsen, a community of 2,000 predominantly black residents located ten miles north of the state capitol building, and Rollins Environmental Services, Inc., which operated the nation's fourth largest hazardous-waste disposal and treatment facility nearby. It also drew widespread attention, in Baton Rouge and elsewhere, after the national environmental justice movement targeted it. For years residents had complained of odors, respiratory problems, eye irritation, and damage to their gardens and property from the Rollins facility. The U.S. Environmental Protection Agency (USEPA) took an interest in the Rollins operation and obtained a consent order to monitor contamination from the land application of wastes. Rollins subsequently installed an incinerator to dispose of the waste. While odors from spreading petrochemical wastes over the land were annoying, residents became frightened by the incinerator when, on August 5, 1985, it malfunctioned and spit out clouds of pollutants over the area, disrupting church services at the Mount Bethel Baptist Church. On another occasion, pollution from the incinerator sent children at a nearby elementary school onto the school grounds vomiting. In fact, by 1986 Rollins had been cited over 100 times by EPA and fined $1.7 million by state environmental regulators. The USEPA, however, was not ready to abandon the disposal site. During this period it waged a war with both the state of Louisiana and local officials in order to allow Rollins to incinerate polychlorinated byphenols (PCBs) sent to Baton Rouge from Superfund sites around the country.

Residents wanted the facility closed, which the state ordered after the school incident. A number of petrochemical firms demanded that the facility remain open, however, since it was critical to their operations, and Rollins quickly obtained a court order to reopen. After residents took their grievances to court in nine separate lawsuits, they received $350,000 in damages (an average of $3,000 per household).[50] The company was fined again in 1991 for illegally receiving and burning radioactive waste. Nevertheless, over resident protests in 1991 the state granted Rollins a ten-year permit to continue.

WHERE WAS LOCAL GOVERNMENT?

A question that naturally arises is: What were the city-parish government and its staff of planners doing as Baton Rouge industrialized and the quality of the environment, in turn, declined precipitously? A search through thousands and thousands of newspaper articles over the period 1950–94 revealed only one instance of local government proposing measures that recognized the hazards posed by development of the chemical industry and offered remedies. That was a 1957 proposal to protect rural land to the north of the city for future industrial use by banning residential development in the designated areas. While the proposal was primarily aimed at meeting manufacturers' needs, it would also have prevented households from coming in close contact with polluting industrial plants.[51] The parish council, however, turned the proposal down, bowing to the demands of a Citizens' Planning and Zoning Committee and a group of industrial realtors who claimed that landowners needed the freedom to sell their property for a variety of potential land uses.

Obviously, much of the work that planners do is technical and not likely to draw newspaper headlines. A review of four major planning documents prepared over this period, however, reveals a similar lack of interest in industrial pollution and its consequences for the quality of life. The 1949 comprehensive plan for the city is completely silent on the issue of industrial pollution, noting that winds blew pollution to the north away from the city center (to the north, of course, was the predominantly black community of Scotlandville, which the city annexed in the 1980s). In 1958 the city-parish government extended zoning regulations to the rural sections of the parish. Industrial zoning was applied to the Exxon refinery and to the heavy industry located to the north, but no attempt was made to adopt industrial performance standards to curb pollution. In 1972 the city updated the comprehensive plan. A committee made up of industry representatives and others advised the planning commission about issues related to industrial pollution. The plan reported that existing safeguards designed to protect plant personnel would be more than adequate to protect surrounding neighborhoods from pollution. The 1991 *Horizon Plan Summary* plan for the parish observed that Baton Rouge had few regulations in place to protect residents from pollution, but it did not propose remedying the situation. Instead, the plan proposed additional industrial development in areas where it already existed and more intense residential development in adjacent neighborhoods.[52]

In contrast to the evidence of inertia by planners and local officials in Baton Rouge and their meek ideas for coping with massive pollution of the city-parish environment, citizens proposed a more daring list of reforms at environmental justice hearings held in Baton Rouge in 1993.[53] Rather than continued industrial expansion, they wanted a moratorium on new plants and on the expansion of existing plants located near minority communities. They called for existing polluting industries to be phased out and replaced by more sustainable, environmentally friendly "green" businesses. They called for relocation of households from "hopelessly polluted" neighborhoods. They wanted compensation for the loss of property values and quality of life, not just in money, but also in expanded public health and other urban services. They wanted to see impact assessments of proposals to site new industrial establishments. And they wanted a voice in decisions about land use that affect their neighborhoods.[54] Many of the suggestions the citizens put forth are strikingly similar to those proposed by reform groups throughout the latter half of the nineteenth century.

CONCLUSION

Progressive reformers at the turn of the century had high hopes that through planning and management of urban development most of the environmental problems wrought by nineteenth-century industrialization could be avoided. This look at the industrial development of Baton Rouge, which began at that time, reveals that the hopes of the reformers never came to pass, at least in lower Mississippi River valley cities such as Baton Rouge. In fact, Baton Rouge residents, particularly blacks and the poor, are exposed to most of the same maladies—air and water pollution, odors, ill health, substandard housing, and chronic unemployment as their nineteenth-century counterparts. The remedies of the nineteenth century—the nuisance suit and pollution control—are still being used and still achieving the same disappointing results for aggrieved residents.

The new ingredients in all of this—city planning—seems to have had virtually no impact. Planners in Baton Rouge never developed measures to counter environmental harms from industrialization. The reasons may have less to do with planners' ignorance of or indifference to the problem and more to do with their apparent lack of political resources in the face of entrenched interests that viewed industrialization and the jobs and income it provided as an unmitigated good. This may be true, but it may be more telling that no attempt seems to have been made to even propose such measures. In this regard, the contentions of historians Maury Klein and Harvey Kantor seem relevant. They argue that planners at an early stage in the development of the profession abandoned reform goals and joined forces with the business elite, preferring small changes at the margin to major changes in the character of urban industrial development. They write, "The best of them challenged the essence of the industrial city, which was the marketplace of tradition. It was a grossly uneven battle, a tragic mismatch of good intention against the leviathan of forces which brought the industrial city into being and which, a century later, have scarcely slowed their momentum."[55]

Barbara Allen

The Popular Geography of Illness in the Industrial Corridor

RECENTLY, THE 130-mile stretch of the Mississippi River between New Orleans and Baton Rouge, sometimes called "Cancer Alley" or the "industrial corridor," has become a focus area for the national environmental justice movement. This corridor along the Mississippi was, at one time, lined with ribbons of sugar-cane plantations; today it has over 130 petroleum processors and chemical plants displacing agriculture as the region's most lucrative industry. Poor people and people of color are now voicing concern over the unjust burden to both their health and their environment that industry has wrought on them with seemingly little reward or restitution.

Prior to the 1960s there were very few chemical plants in this region. From 1964 to 1968, however, petrochemical growth in Louisiana outpaced all other states, including Texas. Rapid growth was attributed to numerous factors. Most important was the availability of the four building blocks of petrochemicals—salt, water, oil, and natural gas—which gave both Texas and Louisiana a competitive advantage in this industry. In Louisiana, however, the chemical industry was also attracted by a lucrative government package. First of all, the state assured there would be no property taxes assessed to support schools and the other functions of local government. Second, the Mississippi River provided a large volume of freshwater unlike other Gulf Coast locations. Companies were aware that with the passage of the 1966 Federal Water Quality Act pollution would soon be a major consideration. Since rivers with a high discharge rate could accommodate larger quantities of pollution, the Mississippi made an ideal location for manufacturers facing the coming regulations. The U.S. Army Corps of Engineers was busy on a new stabilization scheme for the riverbanks, insuring flood protection and excellent docking facilities. The state also promised highways into the area, which included the new Sunshine Bridge, making the small communities accessible to all transportation modes.[1] Finally, the "no politics attitude from the governor's office" and a "hands off" attitude by regulators also lured companies.[2]

One example of riverfront growth during the prosperous 1960s was Geismar in Ascension Parish. The citizens of this rural, largely African American town supported themselves primarily through small farms, fishing, and other cottage industries. The chemical companies that targeted Geismar for development did not consider these residents a valuable resource; instead, they were invisible to the industry. One article describing the new growth phenomenon in Geismar states, "The town has no manpower . . . most of the workers . . . commute."[3] The corporations locating in Geismar during the growth era included Borden Chemicals (1962), Uniroyal Chemical (1962), Rubicon (1965), Shell Chemicals (1967), Allied (1967), Union Texas (1967), and Pioneer (1970).[4] A number of these plants bought their riverfront property from former white plantation owners who then moved, leaving their poorer and minority neighbors behind.[5]

Today the chemical industry in Louisiana annually generates the equivalent of 16,000 pounds of hazardous waste for every citizen in the state. While this Louisiana industry produces only 6.7 percent of the nation's chemicals, it produces 12.5 percent of all hazardous waste generated nationally, not including the hazardous waste that disposal firms import into the state.[6] The region now has dozens of hazardous-waste incinerators and chemical landfills that line the river, many of them exempt from federal hazardous-waste disposal regulations.[7] Chemical plants have released millions of pounds of toxic chemicals directly into the air and water. The landscape has been dramatically transformed by human intervention; the chemical industry has brought vast networks of pipelines and the constant flow of tank cars and chemical transport trucks interrupting the historic fabric of houses, schools, and churches that once dominated this rural region.

The residents along Louisiana's industrial corridor have, in recent years, begun to complain about adverse health effects caused by the chemical industry's pollution. In an effort to persuade government officials that their plight is real, community members, collaborating with scientists, have devised ways to document what they believe are the negative health impacts of the chemical industry in their area. This chapter will look at several scientists in the corridor who have chosen informal or popular epidemiological methods to further the cause of health in areas that surround the chemical industry. Popular epidemiology is the citizen-initiated collection and communication of health information by concerned residents in a community.[8] These methods have proven invaluable in the environmental justice movement as they consolidate and disseminate the knowledge and experience of the community residents and their allies. Community allies, often scientists, are involved in the public translation of science in such a way that it is meaningful to the communities at risk. In contrast, I compare the official government- and corporate-sponsored epidemiological studies. In one case, an epidemiological study was done in one of the communities that previously had initiated its own study; the outcome of the two health studies varied dramatically. Finally, after examining the various methods that have been used for establishing health claims in the industrial corridor, I will discuss some alternative approaches that further the public science agenda, specifically the importance of a community-situated science.

FLORENCE ROBINSON, PARTICIPATORY SCIENCE, AND ZIP CODE STUDIES

Florence Robinson is a professor of biology at Southern University in Baton Rouge. An African American scientist and educator, she was one of the first people in the industrial corridor to devise a systematic method for documenting health claims. Her involvement with environmental issues began in 1989 when Rollins, a hazardous-waste processor near her house in Alsen, applied for a permit to expand its operations.[9] Her neighbors and family had experienced many illnesses such as her son's migraine headaches and her own respiratory problems.[10] The Rollins expansion prompted her to get involved with the Louisiana Environmental Action Network to learn more about the impact of chemicals in her community.

Robinson realized that for the people of Alsen to have a voice in environmental permitting and decision-making they needed to be educated. She explains:

> In my community most of the people that are my age have less than a grade school education because there was no school out here when they were growing up. You're not going to turn people like this into scientists overnight. What they have is something priceless and scientists would do a lot to pay attention to it as they have got incredible common sense.

What was more important to her than making the citizens into scientists was teaching them how science operates. If they understood how science works they would be better able to "debunk a lot of the nonsense that is out there in the name of science," says Robinson. "What I try to do is just teach them the basic process of science, the scientific method," explaining it in everyday language, she adds. Then, when the companies make assertions or health claims, the citizens are better equipped to ask the right questions about the data and the methodologies used. The scientific method consists of observations, followed by classification of observations, leading to a hypothesis, and finally testing that hypothesis.

One of her methods for determining the effects of chemicals in her neighborhood was to establish a health registry. She used what she calls a citizen-survey method of collecting data. She notes that while these informal methods of collecting data may appear ad hoc they are the first step of the scientific method, making observations. Next, following the scientific method, "you raise questions about what is causing the problem, given all the [hazardous-waste] sites in your neighborhood." The scientist then hypothesizes that some of the illnesses may be caused by these sites. It is at the point of hypothesis in the scientific method that communities go to government agencies, hoping that they will be able to investigate officially and test the illness theories of the residents, the fourth and last phase of the scientific method.

But, according to Robinson, traditional science is often insular and unresponsive, much too slow for the people in communities exposed to chemical emissions on a daily basis. She uses other more heterodox scientific methods to gain access to information in her community quickly. One of her many methods is the map survey (figs. 11.1 and 11.2). She distributed the map survey materials in her neighborhood and the neighbors marked illnesses, suspicious smells, and dump sites on the maps and turned them

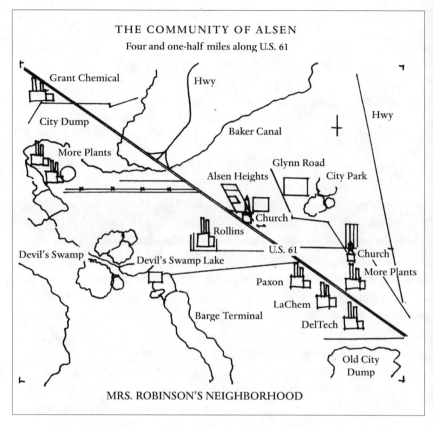

THE COMMUNITY OF ALSEN
Four and one-half miles along U.S. 61

Grant Chemical

Hwy

City Dump

Baker Canal

Hwy

More Plants

Glynn Road

Alsen Heights

City Park

Church

Rollins

Devil's Swamp

U.S. 61

Church

Devil's Swamp Lake

More Plants

Paxon

LaChem

Barge Terminal

DelTech

Old City
Dump

MRS. ROBINSON'S NEIGHBORHOOD

Fig. 11.1. A general survey of the Alsen community just north of Baton Rouge, similar to the maps that Florence Robinson has drawn showing the location of hazardous waste-emitting facilities in her community.

back in. She also went door to door, meeting people in the community and taking down information about health and environmental problems they had experienced. She says about the informal information-gathering process, "I consider it as legitimate as anything that companies can come up with. . . . [i]t may even be more legitimate as industry is biased, the government and the health department are biased [as] it seems to be their role to alleviate the fears of the public."

Robinson uses her maps, which mark various illnesses as well as their proximity to Superfund sites and industry. These visual tools can be easily updated and used in public education lectures as well as exhibits to accom-

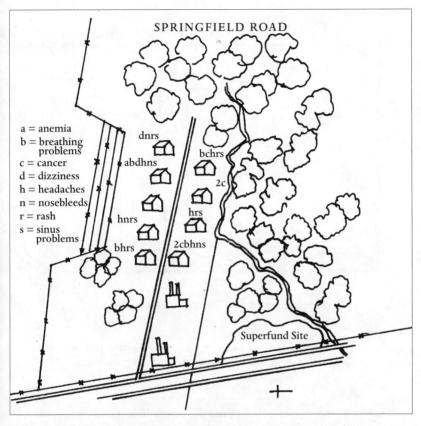

Fig. 11.2. Compilation of one of the neighborhood self-studies that Florence Robinson has done for her community documenting the variety of health problems along one street in Alsen, Louisiana.

pany testimonies before permitting bodies and the state legislature. She pursued her interest in what she terms "environmental racism" by showing a relationship between pollution and race in a series of zip code studies. These studies indicated that there appeared to be a relationship between race and toxic releases in her parish; the "blacker" the zip code area, the higher the quantity of toxic releases (see fig. 11.3).[11] This work has produced empirical evidence of the disproportionate proximity of hazardous waste to minority communities. This study coupled with her other studies on the poor health of those residents living near hazardous industries form the basis of her claim of environmental injustice.

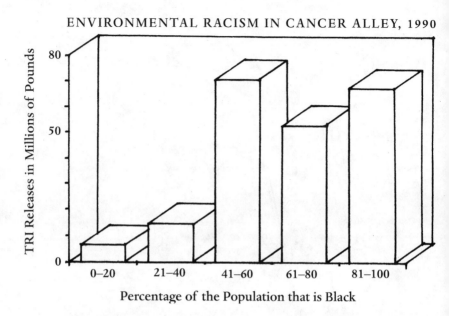

ENVIRONMENTAL RACISM IN CANCER ALLEY, 1990

Percentage of the Population that is Black

Fig. 11.3. Compilation of TRI data done by Florence Robinson showing the relationship between hazardous waste and race in the industrial corridor.

KAY GAUDET AND THE MISCARRIAGE CONTROVERSY

Kay Gaudet and her husband, Chris, moved to St. Gabriel, Louisiana, in 1974, a year after they had both finished pharmacy school. Kay's parents lived there and the couple liked the out-of-doors and the spacious country living. At the time St. Gabriel had a handful of chemical plants, and in 1976, when they opened their own drugstore, the Gaudets even tried to get industry accounts. "We were friendly toward them," says Chris Gaudet. "It was not until 1979 or '80 that the first chemical accident happened that got us involved . . . and nobody in the community was notified of the danger."[12] St. Gabriel was an unincorporated town and the chemical industry continued to grow, with little input from the residents, causing more concern about the possibility of accidents. There are over a dozen chemical companies there today.

The final catalyst for their transformation into environmental activists happened when, according to Kay:

One day we woke up and saw that there was traffic, bumper to bumper, all leaving the area. What had happened was that the facilities had evacuated all their workers out of the area, but they had not bothered to tell the people that live there. . . . We became more concerned as both of our neighbors were elderly and had no way to get out.

Other things concerned them about the operations of the plants nearby. Chris explains:

We used to plant a garden in the back of our house, but we don't anymore as we are 200–300 yards away from Ciba-Geigy and they manufacture herbicides. They were careless in how they packaged their chemicals and thus white powder would escape the plant site. One morning we got up and we looked at our garden and everything was dead.

Soon other signs of problems began to emerge. "One of the top drugs dispensed at our pharmacy was Nolvadex, which is used for breast cancer," says Kay. She made a statement to the paper regarding her concern. Another pharmacist along the industrial corridor contacted her; he, too, was dispensing a lot of the breast-cancer drug. She suggested publicly, but to no avail, that a study be done. Chris adds that the perception is that a disproportionate number of children in the local St. Gabriel school have learning disabilities, but a study was never done by any state agency. At one time Kay even asked the local toxicologist to begin sampling the fatty tissue of people that have died in the area, but no one undertook such a study.

Then, in the mid-1980s, Kay Gaudet received a letter from her sister, living in another state, who had recently suffered a miscarriage. She suspected that the chemical pollution in her area had something to do with her pregnancy loss. This was the first time Kay began to think about the relationship between the miscarriages and toxic pollution. She explained that "between what I knew having a pharmacy and knowing everyone in the community . . . we became aware of a lot of women that had suffered miscarriages." When she made her observations about the number of miscarriages public (in the local newspaper), women she did not know began calling her with their stories of pregnancy loss in St. Gabriel.[13]

At this point, Kay became very concerned about what she perceived as an excessive miscarriage rate in St. Gabriel. She said, "I was keeping a journal and it got kind of scary. One July 4th weekend, we had a long weekend. There had been three women in four days that had miscarried. It occurred to us that when you had one, you had numerous ones, like it was cyclical—we had a feeling that it was related to chemical releases that may have occurred." Kay's informal inquiry led her to believe that in St. Gabriel, a town of 2,100 inhabitants, "approximately one-third of the pregnancies since 1983 had ended in fetal death."[14] This amounted to more that sixty miscarriages by fifty women in three years, twice the national miscarriage rate of 15 percent.[15]

As before, the Gaudets contacted state agencies, asking them to look into what appeared to be an alarming trend. Again, nothing happened. Kay eventually contacted a national organization, the Sierra Club, for help. They agreed to send Kay Gaudet to Washington, D.C., to testify before a congressional hearing on pollution; soon thereafter state officials contacted her. The Louisiana Department of Health and Hospitals paid Tulane University's School of Public Health and Tropical Medicine $100,000 to do a study that was completed in 1989. The investigation included not only the town of St. Gabriel, but the entire eastern portion of Iberville Parish, which is divided into two sections by the Mississippi River. The study found that the overall miscarriage rates for the eastern half of Iberville Parish were between 12.7 percent (documented) and 15.7 percent (documented and undocumented), well within the national norms. The study analyzed 354 women in the eastern portion of Iberville Parish and found that between 1982 and 1987 the women had 327 live births, 54 documented miscarriages, and 15 undocumented miscarriages.[16] Gaudet's informal study, which was done over only a three-year period, looking at fifty women found almost as many miscarriages. Kay Gaudet's study looked only at the small town surrounded by the dozen chemical plants whereas the state study took half the parish, much of which was not even close to a chemical plant, as its study area.

The Tulane report further concluded that there was "no significant association between proximity to industrial site and incidence of miscarriage."[17] It is unclear how the researchers could make this claim. They specifically did not look at the town of St. Gabriel, which is surrounded by industry but instead chose a much larger region to study because the larger population would yield statistically better data. The Tulane study did not correlate pregnancy loss to proximity to industry. Since the larger population of east Iberville Parish was not necessarily exposed to higher levels of industrial pollution, elevated pregnancy loss would not be expected. The state attorney general, William Guste Jr., was not convinced by the study and hired an independent consultant to review the Tulane study. In reviewing the data, Richard Clapp, a consulting epidemiologist from Boston University, found that women living within a half mile of a chemical plant appeared to have a significantly higher rate of miscarriage.[18] Both Guste and Clapp agreed that the study conducted by Tulane was inconclusive at best.

Paul Templet, an LSU environmental studies professor agrees. He says, "[Y]ou can design studies that don't show much."[19] He believes the St. Gabriel miscarriage study done by Tulane is a perfect example of this:

> They did find elevated miscarriage rates among white females. They had trouble interviewing black females—it was a cultural thing. They don't talk about miscarriages as they consider it to be a failure. There was an elevated rate but because of the way scientists do statistics, the result didn't reach the 95 percent confidence level, a standard used by epidemiologists. . . So while they did see an increase in miscarriages among white females, they reported it as "not significant." But the term "not significant" is a statistical term [and] industry picked up the term and started using the word "insignificant" which is not what it means.[20]

Since Templet was secretary of the Department of Environmental Quality (DEQ) during the time the study was completed, he was able to ask the researchers "why they did not do a regression analysis using some surrogates, like distance from the plants to these clusters to see if there is a connection between distance from industry and the miscarriages."[21] They told him that his suggestions were not part of the original design for the study and that no one offered to pay for additional studies. They did admit in the final report that their case ascertainment for black women was low, meaning that the researchers were unable to locate and document many of the early term miscarriages within that group of women. The stillbirth rate (which was considered a miscarriage for purposes of this study) for black women in east Iberville Parish was almost double that for black women in the state, but this was found to be not statistically significant.[22] Templet was disappointed with the study because he felt that without looking for correlations or clusters a scientist cannot begin to ascertain the cause of any abnormality. He felt the study, by encompassing a large enough region to produce proper epidemiological results, missed the miscarriage problems of St. Gabriel.

THE LOUISIANA TUMOR REGISTRY AND
THE POWER OF STATISTICAL DATA

Charity Hospital established Louisiana's first cancer registry in 1974. Transferred to the state Office of Health five years later as the Louisiana Tumor Registry (LTR), it had within several years generated over fifty reports that looked at cancer rates and risks in the state.[23] Consistently, the death rate from cancer in south Louisiana had been very high compared to national statistics. But death rates had been disaggregated from incidence rates for cancer, and no one knew the cancer incidence in the region. State officials felt there was a need to examine the many reports and their methodologies in order to disseminate some conclusive information to the public regarding their risk of developing cancer.

In 1982 the Department of Health and Hospitals hired epidemiologist Vivien Chen as director of the LTR. The LTR literature defines epidemiology as "the study of the distribution and determinants (causes) of diseases in human populations."[24] Many epidemiological studies use inferential statistics to analyze large samples of the population. While this method is not the only experimental design used by epidemiologists, it is the most common. One of the limitations of this technique is that it requires a large sample population, or a very large effect, in order to reach statistically significant results; in other words, the test results should generally reach the 95 percent assurance rate. This means that there is less than one chance out of twenty that the results are due to chance. If the health effect is rare, the sample size must be quite large, for statistical purposes.

It is important to note that there is a methodological concern in the public health assessment community regarding epidemiology. Critics claim that using inferential statistics in epidemiology will create biases that may occlude real community health problems while falsely targeting less important health issues. According to two environmental health scholars, David Ozonoff and Leslie Boden:

> One must first distinguish (as many health officials do not) between public health significance and statistical significance. An increased disease rate can, if true, be of enormous public health significance but not statistically significant in a given study. . . . [If] the number of cases was relatively small, it was not possible to exclude the circumstance that this increase arose by chance, so it was not statistically significant. Conversely, by tak-

ing a large enough sample, even the tiniest difference may be statistically significant, even though it may have little public health importance.[25]

Critics of epidemiology claim that statistical analysis cannot properly be applied to small samples of people or diseases that are relatively rare occurrences. If the health effect is a rare event, as in the case of some types of cancers, an even larger sample of the population is needed to ensure that the occurrence is not simply due to chance. Because of this requirement in inferential epidemiology, the method is inappropriate for looking at individual rural communities along the industrial corridor and will generally yield statistically problematic results because of the small numbers of people and the rarity of some of their diseases.[26] Yet the state board of health persists in using epidemiology as a method to prove that there are no negative effects on the residents due to pollution. This method of assessment is popular with both government and corporate representatives, possibly because it yields results consistent with promoting further chemical industry development in the state.

Richard Couto, a scholar researching the politics of epidemiology, claims:

> The degree of risk to human health does not need to be at statistically significant levels to require political action. The degree of risk does have to be such that a reasonable person would avoid it. Consequently the important political test is not the findings of epidemiologists on the probability of non-randomness of an incidence of illness but the likelihood that a reasonable person, including members of the community of calculation, would take up residence with the community at risk.[27]

Interested in this "reasonable person" standard of why someone would live in a polluted community, I asked Florence Robinson why she continued to live in Alsen, knowing the effects of the chemicals around her. She explained that she had owned her home for fifteen years before she figured out what was happening to her and her neighbors. "You can't sell," says Robinson, "and one lady who moved a few years ago had her house on the market for two years before selling it at a considerable loss." Later she ran into the real estate agent who sold the house. The agent informed her that it was almost impossible to sell houses in the Alsen area because no one will finance the house. The agent said that she had called numerous mortgage companies only to be told "we don't finance houses in that area."[28] According to

Robinson, a new kind of redlining based on environmental hazards was going on in the industrial corridor. Even the banking establishment was leery of owning property located in Alsen.

Industrial communities such as Alsen could, however, benefit from inferential epidemiology if the study was properly designed. Achieving a high degree of statistical certainty in evaluating the occurrence of rare diseases could be achieved by aggregating the population sample differently. Instead of expanding a study to include an entire parish, most of which was not exposed to toxic chemicals, one could expand the study regionally or nationally to include all communities with similar toxic exposure. With TRI data and GIS technology, a study could be designed to look at communities with similar chemical exposures. It would yield results that would be more useful in proving causation and in setting regulatory policy relative to toxic pollution. Such a study would be costly and require cooperation across state and federal agencies as well as the integration of various science methodologies. But if an answer to the question of chemical exposure is important, then prioritizing funding for such an endeavor should follow.

A larger study was eventually funded but not to the benefit of the citizens. In 1988 the Louisiana Chemical Association (LCA) contracted with a private research group to review all of the reports and specifically look at the relationship between cancer incidence and proximity to industry.[29] The report found that there was no correlation between occurrence of this disease and living near a chemical plant. A year later the LCA funded a project by the LSU Medical Center to analyze previously collected cancer data. This project resulted in the publication of "Cancer Incidence in South Louisiana, 1983–1986."[30] The analysis encompassed the thirty-five parishes considered to be south Louisiana, representing 72 percent of the state's population. The study found that, compared to national statistics and "with the exception of lung cancer, residents in South Louisiana have a lower risk of developing most common cancers."[31] The elevated incidence of lung cancer is substantially higher than the norms determined by the North American Association of Central Cancer Registries in all populations except black females.[32] The abnormally high incidence of lung cancer was attributed to smoking. The report further asserted that "the high cancer death rates in South Louisiana are not due to high incidence rates . . . [but that] the major problem appears to be a lack of early detection and limited access to health care."[33] The report also noted that poor nutrition and smoking contributed to both the incidence of and mortality resulting from cancer in the state.

The LTR was careful to deflect any blame for cancer from environmental pollution. They wrote:

> National estimates suggest that occupation accounts for approximately 4–6% of all cancer deaths. Pollution accounts for less that 2%. Epidemiological studies done to date in South Louisiana suggest that the same is true here. . . . Air, water and land pollution have been suggested as major causes of cancer in South Louisiana. The data in this monograph and the many studies conducted to date in South Louisiana do not support this belief.[34]

Placing the blame for cancer and other illnesses on the individual and his or her habits is not new. The LTR envisions its role in the decrease of cancer in the state as one of education; teach the people about the effects of smoking, drinking, and poor nutrition, and their health will dramatically improve. Thus, state intervention is one of information flow, not of environmental regulation, when it comes to most illnesses. This conveniently shifts the responsibility for improving health to the citizens themselves who must learn to stop hurting themselves. In the case of poor and low-income people the problem is magnified as their mortality rate is over three times the mortality rate of middle- and upper-income people combined.[35]

Recently research into the causes of the elevated mortality rates among lower income groups has contradicted the "bad habits" hypothesis. A national eight-year study of over 3,500 individuals published in the *Journal of the American Medical Association* found that even after accounting for smoking, drinking, and obesity, lower-income people still had twice the rate of death from both cancer and heart disease.[36] According to the researchers "health risk behaviors do not explain much of the relationship between economic factors and mortality."[37] To what can this elevated mortality rate be attributed?

First, the study points to differences in exposure to both occupational and environmental hazards and when coupled with a lack of access to medical care and the psychological stress of poverty, are the likely candidates of poor health status.[38] The study concludes that the proper focus of both policy rhetoric and actions should not be on health education at the individual level. Instead, it would be more productive to focus on decreasing environmental hazards, such as toxic exposure, as well as provide more access to medical care before the acute stages of a disease. The fear of toxic exposure, the symptoms of such exposure, and the lack of medical care worsen the stress of poverty.

According to a study done by Paul Templet, Louisiana ranks third in the nation in the percentage of household income spent on health care. On average, a family in Louisiana spends approximately 32 percent of its income on health care. This compares poorly to a state such as New Hampshire, which has a per family expenditure on health care roughly half that of Louisiana.[39]

Albertha Hasten, a welfare and environmental activist living in the industrial corridor, describes what limited health-care access means to her daily life. "The nearest health care facility is ten miles away from me," she says, "I have to hitchhike with my children to get there. So what people around here do is borrow other people's medicine."[40] Hasten lives on a small income and must piece together her health-care needs according to her limited budget. The added expense of transportation from her rural town to the city is beyond her means. Her story echoes the health-care plight of many of the poor residents of the rural industrial corridor.

Thus the LTR's proposal of further education would do very little to decrease the cancer rate along the industrial corridor. Similarly, education would not have a great impact on other diseases such as respiratory illnesses in the corridor. The "lifestyle" excuse for poor health in the region, which is economically depressed, serves only to occlude issues such as poor air and water quality, which the residents have little control over. Lengthy battles to combat pollution often last many years and are themselves very stressful. The LTR's claim that higher mortality rates from smoking and poor diet parallel the "lifestyle" argument made by health department officials in the Ronald Reagan administration. The poor are responsible for their own health problems and what is needed is health education "to counsel minorities [and the poor] to eat better, exercise more, smoke and drink less, be less violent, seek health care earlier for symptoms, and in general be better health-care consumers." Blaming the poor and minority communities for their own elevated mortality rates deflects responsibility from industrial society as a whole for this problem. The conflation of lifestyle choices with lived circumstances is problematic.[41] After all, most of the citizens of Louisiana's industrial corridor did not choose to have chemical plants as their neighbors any more than they choose the type of pollution the plants emit.[42]

By continuing to insist that the possibility of personal habits, like smoking, be ruled out before any investigation into the effects of pollution can be valid insures that toxic exposure can never be studied properly and therefore blamed for health problems. In the case of Love Canal, the New York State Department of Health officials explicitly stated that the possibility of bad habits causing the illnesses had to be ruled out before heath effects could be related to toxic waste.[43] Causation is a heterogeneous mix of elements, both willful and unintentional. State agencies all too often use this as an excuse to do nothing regarding the quality of air and water in the industrial neighborhoods. My research strongly suggests that intervention into corporate environmental practices, not individual habits, is essential to reverse negative health trends among poor and low-income citizens.

The "lifestyle" argument was only one hurdle the citizens of the industrial corridor had to face. Proof of causation in the case of cancer or any other suspected environmentally related disease is difficult to produce. First, the complexity of such a study is immense. The study would need to analyze a group of people living near chemical plants who also may drink, smoke, or have a genetic disposition to cancer. Add to this the fact that chemical contaminants may have a synergistic effect; in other words, several pollutants mixed together may produce something very different than each chemical does individually in the body. Also the length and concentration of exposure, particularly to accidental releases, would have to be factored in. And the gestation period for many diseases, such as cancer, can be as long as twenty to thirty years; the latency period for exposure to low-level pollution may be even longer. Additionally, the many cancer studies done in Louisiana have shown that there exist some substantial differences in both incidence and mortality depending on both sex and race of the individual. The complexity and expense of such a study, even if it could be devised, is such that none has been undertaken to date. The beneficiary of this inaction is the chemical industry, which can unequivocally state that there is no proof their pollution harms neighboring residents.

Take, for example, lung cancer, which is the most common form of cancer in south Louisiana. While 27.2 percent of Louisianans smoke, 2 percent above the national average, lung cancer for white males is 30 percent above the national average and lung cancer for other populations is also elevated. Richard Clapp, an outside consultant, wrote, "The increase in lung cancer in 'Cancer Alley' always intrigued me. If it were due to smoking, I would have expected a similar elevation in other smoking related cancers such as oral cancer and laryngeal cancer, but these were not as striking as the lung cancer excess along the river."[44]

Another anomaly of cancer epidemiology studies in the region is that the risk of incidence of cancer for people living there is at or below the national average, but the mortality rate for cancer in that region is extremely high.[45] The reasons for this high mortality rate given by Chen is late diagnosis combined with poor access to health care, and the fact that lung cancer, the most prevalent form of cancer in the area, kills 85 percent of the time.[46] Clapp has another theory for the high cancer-mortality rate in Louisiana; cancer incidence is underreported by the LTR, a claim that Chen vehemently denies.[47] In the majority of studies performed by the LTR there is an emphasis on incidence, and mortality rate is often dismissed in a paragraph or two. Why the lack of research on cancer mortality in this area since it is unusually high? Are the cancers in the industrial corridor more aggressive? Does additional chemical exposure exacerbate the disease? Would not death from cancer be important to study and have at least as much, if not more, impact on families than simply having the disease in its various forms and phases? Why would the result of this disease (death) not be as important as its incidence?

In the final analysis, epidemiology studies have failed to show what many residents believe, that their constant exposure to a multiplicity of hazardous materials causes cancer. As a matter of fact, the latest epidemiological study of the industrial corridor published in the *Journal of the Louisiana State Medical Society* in April 1998 by Chen and her associates indicates that residents of the chemical corridor have a lower incidence of lung cancer than the state population as a whole.[48] The average citizen reading this study might conclude that it is healthier to live by a chemical plant than to live elsewhere.

Interestingly, cancer is not the disease affecting the largest number of residents. Studies reveal that respiratory disease is a much more common ailment of chemical plant neighbors. In a seven-parish study of low-income residents of the industrial corridor conducted by the University of New Orleans in 1995, residents were asked what kinds of health problems they had that they felt were attributable to pollution. Of the approximately 800 respondents all living within a mile of the river, the most industrialized part of the region, 35 percent of the residents suffer from respiratory problems, 21 percent suffer from allergy problems, and 17 percent suffer from sinus problems; only 9 percent suffer from cancer.[49] Why no investigation of these diseases? Why have all significant investigations of disease in the petrochemical region focused only on cancer?

Patricia Williams, an occupational toxicologist, has the same concern. She says:

> The problem with cancer is that it happens at the end of the spectrum, with people having a long list of diseases before cancer evolves. The other thing is that there are far worse things than cancer that can evolve such as lupus, auto immune disease, and aplastic anemia—they can be life threatening. We have not become environmentally conscious enough to look at those environmental indicators that are not cancer because cancer data can be easily manipulated. It's the first thing that is funded and it's a good look at the problem but it certainly does not tell the whole study. You have a horrific spectrum of problems before you get cancer.[50]

FUTURE PROSPECTS

Can any scientific approach help communities understand the relationship between pollution and health? It seems that what currently exists is a divide. On one side orthodox epidemiology uses complex mathematical modeling tools but provides little insight into the community's health problems. On the other side popular studies show illness clusters but appear anecdotal and as having little rigorous control over the study's content. An approach somewhere between the distant objectivity of epidemiological models and the overly subjective approach of informal local cluster studies is needed.

Two such approaches to understanding the relationship of illness to the environment have been proposed. One takes the individual and his or her own exposure as the center of study, and the other places the community as the focus of study. Together they form a promising, holistic approach to understanding the effects of toxic pollution on the citizens of the industrial corridor.

The first approach is that of Patricia Williams, director of the LSU Medical School Toxicology Outreach Program. She prefers to look at small groups of people; more specifically, she looks at the cellular aspects of toxic exposure with those small groups. She runs a medical surveillance lab that focuses on the early detection of diseases. This, according to Williams, is a new field of medical research that will provide a lens into the origins and first stages of diseases. Her lab currently is studying Grand Bois, a community in south Louisiana, adjacent to a large oilfield waste site and afflicted with numerous diseases and cancers.[51] A well-publicized dispute between a small ethnic (Cajun and Native American) community in south Louisiana and Exxon has centered on whether or not the waste is causing the citizens' illnesses. Williams has collected blood, urine, and other clinical specimens and examined them for parameters that would indicate toxic exposure. When she goes into a small community, she looks for many different illnesses and abnormalities, not just cancer. Her work takes into account the lifestyles and medical background of the people. In this way she can begin to get a picture of everything that is involved in the origin and evolution of a disease. "This approach," says Williams, "is far superior to any numbers game" of the epidemiologist.[52]

She believes that the mathematical models epidemiologists use are too broad to show causation and should only be used as a screening tool to give a possible direction for further research. Epidemiology, according to Williams, should be used in addition to and not in place of, other more localized studies. While the mathematical technique of describing illness works well when looking at acute illnesses over a large population area, it fails when looking at chronic illnesses and less drastic health impacts. "Often," explains Williams, "epidemiology has been over-interpreted outside of the limits" of this discipline. Epidemiologists are often working without in-depth knowledge of the specific diseases they are statistically analyzing. In the case of cancer studies, Williams points to the arbitrary assignment of one-year increments as one example of not understanding the nature of a disease. She explains that "diagnosis and origin of the disease did not occur within the same year" and that different forms of cancer have different development periods that might cause researchers to look further back at a patient's lifestyle and location.[53] Clearly the one-year time frame for understanding the incidence of the disease is not closely related to the functioning of the human body and is not helpful when trying to discover the origins of a disease. Using the one-year increment to produce data that ultimately effects determinations on whether health studies are done, or industrial pollution regulations are increased or enforced, seems arbitrary at best.

Williams has been very critical of the LTR's work in other ways. She believes that the

> manipulation of data into arbitrary regions is unacceptable to accomplish the mission of cancer prevention and use of the LTR data for research. Regional rates dilute the data. It does not take a scientist to realize that a multiple parish region with one parish reporting high incidence rates will be mathematically blended into a region with no increased incidence of cancer. Is this a deliberate attempt to mask important cancer data?[54]

According to Williams, government officials and elected representatives have tried to acquire access to the LTR's raw data to evaluate the cancer statistics at the single parish and local level; their requests were all denied even though it is public information.[55] The veil of secrecy that surrounds the LTR does little to instill citizen confidence in either methods or ultimate goals of the agency.

Williams respects community-generated health studies and, although she has never met Kay Gaudet, she heard about her miscarriage study. She felt that Gaudet made some excellent observations that should have initiated a detailed study of reproductive problems in the St. Gabriel area. Miscarriage, however, is the least sensitive indicator of reproductive problems as about 15 percent of all pregnancies end this way normally. Instead, Williams recommends looking at indicators such as the number of birth defects, the number of children born that live less than thirty days to a year, the number of premature births, and the number of ectopic and molar pregnancies. She has used this methodology and found "tremendous differences between the control population and the exposed population."[56]

Studying the effects of pollutants on the fetal phase of human development makes sense to Williams since "embryonal cells, by their very nature, are highly proliferative (undergo extensive cellular division) and chemically mediated (respond to chemical stimuli)." She also notes that residual embryonal cells remain in the body until puberty and, "if triggered into cellular division after birth, can form tumors," such as rhabdomyosarcoma and neuroblastoma.[57] For this reason, studying childhood cancers may be another conclusive way to understand the effects of toxic exposure in humans.

It is important to note that while childhood cancer is relatively rare, its rate of increase has been about 1 percent per year, making it the most common fatal childhood disease, according to the National Cancer Institute. The rarity of cancer among this age group makes it difficult to discern both trends and causes using epidemiological statistical analysis, as this type of analysis works best for determining common illnesses among large population samples. Health specialists advising the USEPA believe that toxins in the air, water, and food of children are one of the prime suspects in this alarming trend.[58] Says the USEPA's children's health advisor, Dr. Philip Landrigan, a pediatrician who directs the division of environmental medicine at Mount Sinai School of Medicine in New York: "The increases (in childhood cancer) are too rapid to reflect genetic changes, and better diagnostic detection is not a likely explanation. The strong probability exists that environmental factors are playing a role."[59]

One specific form of cancer, rhabdomyosarcoma, a highly malignant, soft-tissue tumor, is diagnosed in about 200 children in the United States each year, which is, statistically speaking, one in every 250,000 children. In the early 1990s three children in the predominantly rural and heavily industrialized Ascension Parish were diagnosed with the disease within a fourteen-month period. The children lived within a six-mile radius of one another. While this was alarming, it was not the first instance of a childhood cancer cluster in the industrial corridor. Several years earlier, in another heavily industrialized parish (East Baton Rouge), there had been a cluster of four children diagnosed with rhabdomyosarcoma. The children lived in Zachary, a few miles from Alsen and the various industry and Superfund sites that populate Florence Robinson's neighborhood. The oncologist treating the four children requested a study on what appeared to be a concentration of such a rare illness in a very short period of time. The four children died and no study was ever done.[60]

There are other indicators that the diagnosis of childhood cancer and other catastrophic childhood illnesses may be growing along the industrial corridor. One of the major treatment centers for this disease and other catastrophic childhood diseases is the St. Jude Children's Research Hospital in Memphis, Tennessee. Patients at this hospital come from nearly every state in the nation as well as from over fifty foreign countries. Since the research hospital opened in 1962, Louisiana children with cancer accounted for nearly 13 percent of all the patients treated at the facility. In 1996, of the 567 children from Louisiana on "active status," 40 percent came from a four-parish area within the industrial corridor.[61] Some health officials explain that these staggering numbers are due to "referral patterns," meaning that physicians in this area tend to send children to St. Jude rather than other cancer-treatment facilities such as M. D. Anderson in Houston.[62] Despite the growing concern regarding the prevalence of this disease, the LTR's most recently published monograph, *Cancer Incidence in Louisiana, 1988–1992*, authored by Vivien Chen and others, fails to list important pediatric cancers such as rhabdomyosarcomas and neuroblastomas. This omission was justified by Chen, who claimed that she did not have the data available to her.[63]

Patricia Williams believes that the numbers of these rare childhood cancers warrant an investigation before dismissing them. She explains that "given the rare incidence of these tumors and their proliferative nature upon chemical stimulation, they may serve as environmental indicators when occurring in clusters in communities that are adversely impacted by toxic exposures."[64] Although Williams's micro-approach of beginning with the body and working inward to a cellular level holds promise at this point, many communities and community-oriented scientists and physicians in the chemical corridor have little idea of the variety of pollutants affecting individuals in the region.

This is where the second approach to understanding the effect of toxins on communities can be helpful. Determining the effects of pollution on nearby communities takes looking at the proximity of the community to industrial pollution and mapping that information onto the health data from the community. Charles Flanagan, a geography graduate student at Louisiana State University, proposes a more location-specific approach to toxic-emission documentation, particularly in areas that have been involved in environmental justice struggles. He proposes that, for purposes of pollution studies, instead of making the chemical plant the center of the map and drawing one-, two-, and three-mile rings radiating out from the facility as in permit application maps submitted by industry, the community to be studied needs to be placed at the center of the map (see fig. 11.4). That way the focus of any community pollution study would be on the total impact of all the emission sources on that community. Says Flanagan:

> I think we need to take the community as the focus of analysis. In this way, we can begin to develop a picture of the potential cumulative impact of any new facility. . . . We need to construct the size of the study area around the community in a manner that relates to toxic air dispersions [having to do with wind speed and direction] where there have been actual accidents.[65]

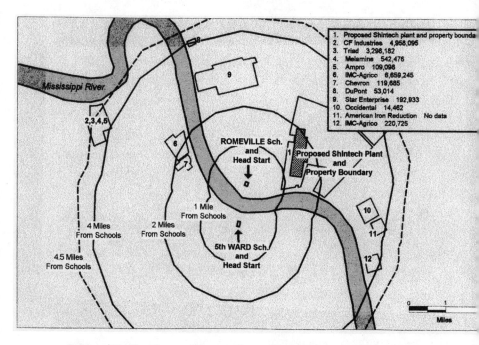

1. Proposed Shintech plant and property bounda
2. CF Industries   4,958,095
3. Triad   3,298,182
4. Melamine   542,476
5. Ampro   109,098
6. IMC-Agrico   6,659,245
7. Chevron   119,685
8. DuPont   53,014
9. Star Enterprise   192,933
10. Occidental   14,462
11. American Iron Reduction   No data
12. IMC-Agrico   220,725

Mississippi River

2,3,4,5

ROMEVILLE Sch.
and
Head Start

Proposed Shintech Plant
and
Property Boundary

1 Mile
From Schools

4 Miles
From Schools

2 Miles
From Schools

4.5 Miles
From Schools

5th WARD Sch.
and
Head Start

Miles

Fig. 11.4. Charles Flanagan's map of a proposed Shintech chemical plant site in Romeville/Convent area along the industrial corridor. His community-focused research places the town at the center of analysis, a first step in showing cumulative exposure to a mixture of hazardous emissions.

In this way TRI data and other important information on toxic material can be presented in a way that is meaningful to the community at risk. This type of pollution study is exposure-driven and allows community-oriented heath studies to focus on those populations that have been placed at risk rather than relying on statistically driven multiparish health data.[66] In this way "exposed" and "control" populations can reflect actual events and lived experience important in collecting health data that is community-oriented.

By developing a geographic model that places the community and its citizens at the center, Flanagan's methodology creatively intersects traditional disciplinary science and the situated politics of the industrial communities. In the same way that Williams desires to see diseases close up and learn from this complex interplay of variables, Flanagan desires to analyze the local community's exposure more thoroughly. By placing the specific resident or community at the center of an investigation, science is constructed around what is happening to the people, rather than people being constructed to fit mathematical scientific models. It is hoped that, with the help of scientist-allies, yet another human transformation of this river region will take place; a reclamation of the residents' environment for future generations. I believe that by pursuing this type of citizen-situated science with methodologies that begin with people's experiences, answers to the community's questions about toxic exposure can be found. This knowledge will provide a powerful tool for changing the lower Mississippi River industrial corridor into a more livable place, where, in the spirit of democracy, citizens make the decisions regarding the quality of life they envision for their families and their communities.

Fish Diversity in a Heavily Industrialized Stretch of the Lower Mississippi River

THE MISSISSIPPI River Basin drains 41 percent of the coterminous United States and receives discharges and runoff over this entire area from industry, municipal sewage-treatment facilities, and agricultural lands. Nowhere is the assault on river water quality greater than in the final 300 miles of the river in Louisiana. Here, over 350 industrial and municipal facilities line the river. Roughly 175 of them discharge waste water into the river.[1] The 130 miles of river between Baton Rouge to New Orleans—the so-called industrial corridor—is an area of concentrated industrial activity, with 126 petrochemical plants and seven oil refineries. Toxic chemical discharges to this stretch of the river amounted to roughly 150 million pounds in 1992.[2]

Through a long history of channel modification, humans have dramatically altered the natural character of the river, forcing it to remain within its banks and follow a fixed course. As a consequence, the river's delta is no longer growing, and the Gulf of Mexico is reclaiming coastal marshes vital to southeastern Louisiana at an alarming pace. Due to the anthropogenic changes, the river's depth, flow regime, and sediment load are less variable than in the past. Habitats within the river ecosystem have also changed dramatically as a result of human modification.[3]

Despite this dire picture of environmental disturbance, the lower industrial corridor of the river continues to support a diverse array of fishes, with as many as 143 freshwater and marine species. In this chapter I describe how the character of the lower Mississippi River has changed as a result of human activity over the last two centuries. I describe the fish community of the lower 300 miles of the Mississippi River, summarizing past accounts in the fish literature and collections archived in the Tulane University Museum of Natural History. I explain why such a modified stretch of river supports such a diverse fish community, how the fish community is likely being affected by human modification of the river, and why our understanding of how the community is being impacted is so limited. I conclude with an overview of a research program designed to assess the ecological impacts of anthropogenic changes to the lower Mississippi River ecosystem, using currently available information as a baseline.

HUMAN MODIFICATION OF THE LOWER MISSISSIPPI RIVER ECOSYSTEM

A basic characteristic of the human species is the ability to transform landscapes and watersheds to suit its needs. The history of human modification of the Mississippi River certainly did not begin with western European colonization of the North American continent.[4] However, the pace and degree of technological sophistication involved in modifying the river has been greatly accelerated since this time. Starting with levee heightening in colonial times, and continuing through dredging and snagging activities of the early to middle 1800s (removal of submerged trees and tree stumps), meander cutoffs and lock and dam construction of the early 1900s, and construction of dike-fields, revetments and other "training" structures in more recent times, modern humans have systematically tamed the river, forcing it to remain within its banks and follow a set course.

Local, state, and federal government agencies implemented these changes to reduce flooding and eliminate hazards to navigation. Unfortunately, they altered the natural dynamics of the river ecosystem in the process. Dams on the upper Mississippi River and all of the major tributaries of the lower Mississippi River have changed the rate of water flow and the nature of sediment transport and delivery. The river no longer switches channels within its alluvial valley, a natural process that occurred periodically to reduce impediments to flow. The process of delta formation, which was responsible for depositing all of the land in Louisiana south and east of Baton Rouge (the so-called deltaic plain), is now severely constrained. As a consequence, older deltas are subsiding and the sea is reclaiming coastal marsh habitat at an alarming pace. Coastal marshes in Louisiana, as elsewhere, play an important role as nursery grounds for fish and shellfish. The health of this ecosystem is dependent on periodic freshening and sedimentation by the river to counteract the effects of subsidence. In its more dynamic natural state, the river swept back and forth across its deltaic plain, filling in old subsided deltas and rejuvenating surrounding marsh habitat. The present delta now extends out to near the edge of the continental shelf, causing the river to release its sediments and nutrients to deep waters of the Gulf of Mexico. The nutrients, originating as fertilizer applied to crop lands of the upper Midwest, are causing excessive algal growth in the Gulf of Mexico. As the algae sinks, it dies, robbing the waters of the Gulf of Mexico of oxygen and affecting fish and shellfish on the sea floor.[5]

As a consequence of the heavy industrial discharges the lower Mississippi River receives, sediments, water, and fish in the river now have detectable residues of many toxic substances. Researchers are sounding the alarm about certain classes of organic contaminants found in municipal and industrial discharges (plasticizers, pesticides such as DDT, and a common class of industrial and household surfactants commonly referred to as alkylphenols), because experiments show that these substances mimic estrogen and disrupt sexual and other estrogen-related development in vertebrates at very low concentrations.[6]

Industrial and municipal discharges to the lower Mississippi River are affecting water quality in other ways. Between 1958 and 1987 significant increases in strong acid ions and decreases in pH and alkalinity occurred in the lower 200 miles of the Mississippi River.[7] Researchers correlated the increased acidity with discharges of acids and acid-forming toxic wastes. Other work suggests that discharges from the many municipal sewage-treatment plants that line the lower Mississippi River are responsible for the high fecal coliform counts recently recorded in lower reaches of the river.[8] Though a natural component of the intestinal flora of humans and others animals, coliform bacteria contaminate the flesh of fish and shellfish and render water bodies unfit for human recreational contact at concentrations above 8,000 organisms per gallon of water. Between 1982 and 1992 fecal coliform counts exceeded this minimum standard at two-thirds of the Mississippi River stations monitored by the U.S. Geological Survey and the U.S. Environmental Protection Agency (USEPA), including five stations between Baton Rouge and Venice, Louisiana.

Between 1990 and 1994 the Louisiana Department of Environmental Quality (LADEQ) screened 154 samples of fish and shellfish collected from the lower Mississippi River and its adjacent floodplain for a standard list of pollutants given priority because of their toxicity to humans. The LADEQ used the information gathered from this effort to estimate the risks to humans of consuming fish and shellfish from the river.[9] The study found that lifetime cancer risks from consumption of three meals or more of fish per month exceeded advisory levels established by the Louisiana Department of Health and Hospitals. The researchers attributed the highest risks with consumption of bottom-feeding fish such as carp and buffalo, and predators such as gar. The compound found to have the highest single-contaminant risk was hexachlorobenzene (HCB), a selective fungicide, and a commonly used starting material and byproduct of the chemical manufacture industry. The LADEQ researchers detected HCB in 130 of 154 samples analyzed (84 percent), and had an average concentration of 0.032 parts per million (ppm). This is higher than the incidence and the average fish tissue levels of HCB found in a backswamp of the river that receives contaminated effluents from an USEPA Superfund site.[10] Though the concentrations were not high, the levels of mercury and polychlorinated byphenols (PCBs) detected in fish tissues also caused some concern because of the toxicity of these substances. The LADEQ study did not attempt to evaluate the ecological risks any of the contaminants pose to the aquatic life.

Scientists have documented egg (yolk) proteins and depressed serum testosterone—normally characteristic of female fish—in the blood of feral male carp (*Cyprinus carpio*) from streams in the upper Mississippi River basin that receive sewage-treatment effluent and agricultural runoff.[11] They attributed this feminization phenomenon to estrogen-mimicking contaminants in the effluents. It is reasonable to suspect that fish in the lower terminus of the Mississippi River have similar problems of sexual expression, because of the high volume of sewage, industrial, and agricultural discharges the river receives from throughout its basin.

FISHES OF THE INDUSTRIAL CORRIDOR

To understand how the fish community in the lower corridor of the Mississippi River has responded to changes in the river's environmental quality, one must compare quantitative data on the community's composition at different points in time leading to the present. Fishes are the best known and most extensively studied organisms in the lower Mississippi River. However, most of the early studies are of a taxonomic or biogeographic nature. These studies report occurrences of particular species but tell us little about sampling location, capture methods, numbers of specimens caught, or museum documentation.[12]

One of the better early reports documents a collection of fishes from the Mississippi River in the vicinity of New Orleans in the winter of 1882–83.[13] The report provides a reasonably precise sampling location and a list of the species taken, along with the museum number each of the samples is cataloged under (useful for verifying species identifications), the number of individuals archived (presumably the number captured), and characters used in the identifications. The report also gives us a general sense of which fish species were present in the river in the vicinity of New Orleans during the winter of 1882-83 (27 species). However, the account is of limited usefulness beyond this because it provides no information on gear type(s) used, amount of sampling effort applied, or habitats sampled. As recent studies on small streams demonstrate, species representation and abundance relationships can vary greatly from one sample to the next, even within habitats, depending on factors like time of day, habitat type sampled, and amount of effort applied. The problem of repeatability of results is compounded in large streams because sampling efficiency decreases with increasing depth and water body size.

What is needed to characterize changes adequately in the fish community in a stream as large as the lower Mississippi River are intensive sampling efforts, involving multiple samples, taken from fixed stations, over one or more annual cycles, using multiple gear types, and with corresponding data on the amount of time, habitat, and water conditions at each sampling. Unfortunately, natural resource agencies have conducted a very limited number of surveys of this magnitude, and those were conducted only in the last three decades. Thus, available information will not permit a comprehensive analysis of how fishes in the lower Mississippi River have responded to changes in river environmental conditions. What the available information will permit is a description of the fish community present in the river during the past thirty years, an explanation of why the community is so diverse, and a discussion—albeit without supporting data—of how this community likely differs from the community that existed when the river was more pristine.

The discussion that follows is based on three main information sources and scattered reports from the ichthyological literature. The principal source of information is records in the fish collection of the Tulane University Museum of Natural History. Archived in the museum are 194 collections from the lower corridor of the Mississippi River. The collections span a period of forty-three years (1956-98) and contain a total of 139,000 specimens representing 133 species in 39 families. Included are collections from three monthly fish-monitoring surveys: a survey involving multiple sites in the vicinity of Baton Rouge and St. Francisville (river mile 240-290), conducted from October 1976 to December 1979; a survey conducted from January 1983 to August 1986 in the vicinity of English Turn Bend (river mile 78); and a survey conducted from September 1986 to December 1988 in the vicinity of Fort Jackson (river mile 19.7) (see fig. 12.1). The surveys involved standardized gear (seines and trammel nets) and sampling effort. Royal D. Suttkus, the former Curator of Fishes and founder of the Tulane University Museum of Natural History, supervised all of the survey collections and most of the nonsurvey collections.

A second source of information is a survey of the fishes of the Delta National Wildlife Refuge, very near where the river meets the Gulf of Mexico.[14] This survey, conducted from August 1963 to February 1965, also involved standardized sampling gear and technique (boat electrofishing, trawling, rotenone). The survey yielded a total of 79 fish species in 33 families.

The third principal source is a survey conducted by V. A. Guillory from February 1972 to December 1973 in the vicinity of St. Francisville (river mile 255-270).[15] The survey involved monthly samples taken with a combination of seines and trammel nets of specified sizes, with specified amounts of sampling effort. In addition to reporting on all fishes collected in the 1972-73 survey (22,509 specimens representing 78 species in 30 families), Guillory documented occurrences of 43 additional species from earlier published reports, records in the fish collection at University of Louisiana, Monroe, and unpublished reports of environmental-monitoring surveys. Guillory also

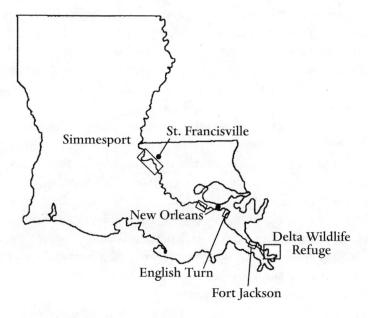

Figure 12.1. Map of Louisiana showing the lower industrial corridor of the Mississippi River, locations along the river referred to in the text, and (boxes) areas of the river where fish surveys reported in this paper were conducted.

related occurrence of fishes in the river and its adjacent floodplain to factors such as season and river discharge.

Table 12.1 is a list of all of the fishes whose occurrence in the lower industrialized corridor of the Mississippi River is documented by specimens archived in natural history collections (principally the Tulane Museum of Natural History). Species names follow the American Fisheries Society Checklist of Common and Scientific Names.[16] Superscripted letters following the species name denote information source. Under "habitat," species are categorized as either primary freshwater (PF), secondary freshwater (SF), diadromous (D), or euryhaline marine (EM). Primary freshwater fishes are species that depend on freshwater throughout life. These species are intolerant of even low levels of salinity (salt concentration) and thus are never found in marine habitats. Secondary freshwater fishes are dependent on freshwater, but are capable of tolerating low levels of salinity for short periods of time. Diadromous fishes are species that move between salt- and freshwater habitats in the course of completing their life cycle. Finally, euryhaline marine fishes are species that are primarily marine, but have a wide range of salinity tolerance. These species are capable of surviving in freshwater for long periods of time.

In all, 134 fish species are known from the industrialized reach of the lower Mississippi River. Virtually all the major groups of fishes found in freshwater around the world are represented. Most of the fishes in the river (76 of the 134 species) are "primary freshwater fishes" species that are restricted to freshwater throughout life. Among these are a species of parasitic lamprey (a primitive jawless fish that superficially resembles an eel),[17] two species of riverine sturgeons of genus *Scaphirhynchus,* the paddlefish (one of only two extant species from this group in the world), four species of gars, the world's only living bowfin, two species of mooneyes, one pike, twenty-three species of minnows, five species of suckers, five species of freshwater catfish, the pirate perch, one freshwater silverside, four species of freshwater killifishes, a pigmy sunfish, one freshwater temperate bass, fourteen sunfish, nine perch, and the freshwater drum. These species dominate the fish community in all but the lowest reach of the river.

The remaining species are capable of migrating between the river and surrounding saltwater environments. One group—the so-called secondary freshwater fishes—favors freshwater but is capable of withstanding low to moderate salinities for short periods of time. Six secondary freshwater fishes are found in the river. Some of these species are wide-ranging, whereas others are limited to marsh habitats in the lower reaches of the river.

A second group of migratory fishes found in the river is the so-called diadromous fishes. These fish move between salt- and freshwater over long time spans (annually or once in a lifetime), typically using one environment for breeding and early development and the other as feeding or resting grounds. Six of the species found in the lower Mississippi River fall into this category. Five of these species use the river as a spawning area after spending most of their time growing and maturing at sea (so-called anadromous fishes). The sixth (the American eel) migrates out to sea to spawn after spending most of its life in freshwater (a catadromous species). When diadromous fishes are in the freshwater phase of their life cycle, their movements within freshwater are limited only by dams and other obstructions. Thus, they occur throughout much of the lower Mississippi River (below St. Louis, Missouri), and in the lower courses of a number of the river's major tributaries, but they are rare above impoundments on these systems.

Table 12.1. Fishes of the Industrial Reach of the Lower Mississippi River

Family	Species name	Common name	Habitat
Dasyatidae—stingrays (<0.01%)	*Dasyatis sabina*	Atlantic stingray	EM
Acipenseridae—sturgeons (<0.01%)	*Scaphirhynchus albus*	Pallid sturgeon	PF
	Scaphirhynchus platorynchus [a]	Shovelnose sturgeon	PF
Lepisosteidae—gars (0.11%)	*Lepisosteus oculatus* [b,c]	Spotted gar	PF
	Lepisosteus osseus [c]	Longnose gar	PF
	Lepisosteus platostomus [b,c]	Shortnose gar	PF
	Lepisosteus spatula [b,c]	Alligator gar	SF
Amiidae—bowfin (0.13%)	*Amia calva* [b,c,d]	Bowfin	PF
Hiodontidae—mooneyes (0.18%)	*Hiodon alosoides* [b,c]	Goldeye	PF
	Hiodon tergisus [c]	Mooneye	PF
Elopidae—tarpons (<0.01%)	*Elops saurus* [b]	Ladyfish	EM
Anguillidae—freshwater eels (0.09%)	*Anguilla rostrata* [b,c,d]	American eel	D
Ophichthidae—snake eels (<0.01%)	*Myrophis punctatus*	Speckled worm eel	EM
Clupeidae—herrings (20.6%)	*Alosa chrysochloris* [b,c]	Skipjack herring	D
	Brevoortia patronus [b]	Gulf menhaden	EM
	Dorosoma cepedianum [b,c,d]	Gizzard shad	D
	Dorosoma petenense [b,c]	Threadfin shad	D
Engraulidae—anchovies (8.22%)	*Anchoa mitchilli* [b,c]	Bay anchovy	EM
Cyprinidae—minnows (17.83%)	*Cyprinus carpio* [c]	Common carp	PF
	Cyprinella camura [c]	Bluntface shiner	PF
	Cyprinella venusta [b,c]	Blacktail shiner	PF
	Lythrurus fumeus [c]	Ribbon shiner	PF
	Hybognathus nuchalis [c]	Mississippi silvery minnow	PF
	Macrhybopsis aestivalis [c]	Speckled chub	PF
	Macrhybopsis gelida [c]	Sturgeon chub	PF
	Macrhybopsis storeriana [c]	Silver chub	PF
	Notemigonus crysoleucas [b,c,d]	Golden shiner	PF
	Notropis amnis	Pallid shiner	PF
	Notropis atherinoides [c]	Emerald shiner	PF
	Notropis blennius [c]	River shiner	PF
	Notropis buchanani [c]	Ghost shiner	PF
	Notropis longirostris [c]	Longnose shiner	PF
	Notropis lutrensis	Red shiner	PF
	Notropis maculatus [c]	Taillight shiner	PF

Table 12.1. *(continued)*

Family	Species name	Common name	Habitat
	Notropis sabinae	Sabine shiner	PF
	Notropis shumardi[c]	Silverband shiner	PF
	Notropis texanus[c]	Weed shiner	PF
	Notropis volucellus[c]	Mimic shiner	PF
	Opsopoeodus emiliae	Pugnose minnow	PF
	Pimephales notatus[c]	Bluntnose minnow	PF
	Pimephales vigilax[c]	Bullhead minnow	PF
Catostomidae—suckers (0.97%)	*Carpiodes carpio*[b,c,d]	River carpsucker	PF
	Cycleptus elongatus[c]	Blue sucker	PF
	Erimyzon sucetta[a,d]	Lake chubsucker	PF
	Ictiobus bubalus[b,c,d]	Smallmouth buffalo	PF
	Ictiobus cyprinellus[b,c,d]	Bigmouth buffalo	PF
	Ictiobus niger[c]	Black buffalo	PF
Ictaluridae—N.A. catfishes (1.38%)	*Ameiurus melas*[c]	Black bullhead	PF
	Ameiurus natalis[b,c,d]	Yellow bullhead	PF
	Ictalurus furcatus[b,c,d]	Blue catfish	PF
	Ictalurus punctatus[b,c,d]	Channel catfish	PF
	Pylodictis olivaris[c]	Flathead catfish	PF
Ariidae—sea catfishes (0.02%)	*Arius felis*[b]	Hardhead catfish	EM
	Bagre marinus[b]	Gafftopsail catfish	EM
Esocidae—pikes (<0.01%)	*Esox americanus*[c,d]	Grass pickerel	PF
Osmeridae—smelts (<0.01%)	*Osmerus mordax*	Rainbow smelt	D
Aphredoderidae—pirate perch (<0.01%)	*Aphredoderus sayanus*[c]	Pirate perch	PF
Gadidae—cods (<0.01%)	*Urophycis floridana*	Southern hake	EM
Belonidae—needlefishes (0.06%)	*Strongylura marina*[b,c]	Atlantic needlefish	EM
Cyprinodontidae—killifishes (0.71%)	*Cyprinodon variegatus*[b]	Sheepshead minnow	EM
	Adinia xenica	Diamond killifish	EM
	Fundulus blairae[b,c]	Starhead topminnow	PF
	Fundulus chrysotus[c,d]	Golden topminnow	PF
	Fundulus grandis[b]	Gulf killifish	EM
	Fundulus jenkinsi[b]	Saltmarsh topminnow	EM
	Fundulus notatus	Blackstripe topminnow	PF
	Fundulus olivaceus	Blackspotted topminnow	PF
	Fundulus pulvereus[d]	Bayou killifish	SF
	Fundulus similis	Longnose killifish	EM
	Lucania parva[b]	Rainwater killifish	SF

Table 12.1. *(continued)*

Family	Species name	Common name	Habitat
Poeciliidae—livebearers	*Gambusia affinis*[b,c,d]	Western mosquitofish	SF
(17.95%)	*Heterandria formosa*	Least killifish	SF
	Poecilia latipinna[b,c,d]	Sailfin molly	SF
Syngnathidae—pipefishes	*Syngnathus scovelli*[b,c]	Gulf pipefish	EM
(<0.01%)			
Triglidae—searobins (<0.01%)	*Prionotus tribulus*	Bighead searobin	EM
Moronidae—temperate basses	*Morone chrysops*[b,c]	White bass	PF
(0.81%)	*Morone mississippiensis*[b]	Yellow bass	PF
	Morone saxatilis	Striped bass	PF
Elassomatidae—pigmysunfishes	*Elassoma zonatum*[c,d]	Banded pigmy sunfish	PF
(<0.01%)			
Centrarchidae—sunfishes	*Centrarchus macropterus*[c]	Flyer	PF
(2.47%)	*Lepomis cyanellus*[b,c,d]	Green sunfish	PF
	Lepomis gulosus[c,d]	Warmouth	PF
	Lepomis humilis[b,c]	Orange-spotted sunfish	PF
	Lepomis macrochirus[b,c,d]	Bluegill	PF
	Lepomis marginatus[b,c]	Dollar sunfish	PF
	Lepomis megalotis[c]	Longear sunfish	PF
	Lepomis microlophus[b,c]	Redear sunfish	PF
	Lepomis miniatus[b,c,d]	Spotted sunfish	PF
	Lepomis symmetricus[c,d]	Bantam sunfish	PF
	Micropterus punctulatus[c]	Spotted bass	PF
	Micropterus salmoides[b,c,d]	Largemouth bass	PF
	Pomoxis annularis[b,c]	White crappie	PF
	Pomoxis nigromaculatus[b,c,d]	Black crappie	PF
Percidae—perches (0.05%)	*Etheostoma asprigene*	Mud darter	PF
	Etheostoma chlorosomum[c]	Bluntnose darter	PF
	Etheostoma gracile[c]	Slough darter	PF
	Etheostoma proeliare[c]	Cypress darter	PF
	Percina caprodes[c]	Logperch	PF
	Percina sciera[c]	Dusky darter	PF
	Percina shumardi[c]	River darter	PF
	Percina vigil[c]	Saddleback darter	PF
	Stizostedion canadense[c]	Sauger	PF
Carangidae—jacks (0.11%)	*Caranx hippos*[b]	Crevalle jack	EM
	Oligoplites saurus[b]	Leatherjack	EM
Lutjanidae—snappers	*Lutjanus griseus*[b]	Gray snapper	EM
(<0.01%)			

Table 12.1. (continued)

Family	Species name	Common name	Habitat
Atherinidae—silversides	*Labidesthes sicculus*[c]	Brook silverside	PF
(6.4%)	*Membras martinica*[b]	Rough silverside	EM
	Menidia audens	Mississippi silverside	PF
	Menidia beryllina[b,c]	Tidewater silverside	EM
Gerreidae—mojarras (0.02%)	*Eucinostomus argenteus*	Spotfin mojarra	EM
Sciaenidae—drums (0.91%)	*Aplodinotus grunniens*[b,c,d]	Freshwater drum	PF
	Bairdiella chrysoura[b]	Silver perch	EM
	Cynoscion arenarius[b]	Sand seatrout	EM
	Cynoscion nebulosus[b]	Spotted seatrout	EM
	Leiostomus xanthurus[b]	Spot	EM
	Micropogonias undulatus[b]	Atlantic croaker	EM
	Pogonias cromis[b]	Black drum	EM
	Sciaenops ocellatus[b]	Red drum	EM
Mugilidae—mullets (0.45%)	*Agonostomus monticola*	Mountain mullet	EM
	Mugil cephalus[b,c]	Striped mullet	EM
	Mugil curema	White mullet	EM
Eleotridae—sleepers (1.41%)	*Dormitator maculatus*[b]	Fat sleeper	EM
	Eleotris pisonis[b]	Spinycheek sleeper	EM
	Gobiomorus dormitor	Bigmouth sleeper	EM
Gobiidae—gobies (18.63%)	*Evorthodus lyricus*[b]	Lyre goby	EM
	Gobiodes broussoneti[b]	Violet goby	EM
	Gobionellus boleosoma[b]	Darter goby	EM
	Gobionellus oceanicus[b]	Highfin goby	EM
	Gobionellus shufeldti[d]	Freshwater goby	EM
	Gobiosoma bosci[b]	Naked goby	EM
Scomberidae—mackerels	*Scomberomorus maculatus*	Spanish mackerel	EM
(<0.01%)			
Bothidae—lefteye flounders	*Citharichthys marcrops*	Spotted whiff	EM
(0.42%)	*Citharichthys spilopterus*[b]	Bay whiff	EM
	Paralichthys lethostigma[b]	Southern flounder	EM
Achiridae—soles (0.02%)	*Trinectes maculatus*[b]	Hogchoker	EM

Source: Based on record in the Tulane Fish Collection and other sources noted below.

[a]Not based on Tulane records

[b]D. S. Jordan, "List of Fishes Collected in the Vicinity of New Orleans by R. W. Shufeldt, U.S.A.," *Proceedings U.S. National Museum* 7 (1884): 318–22.

[c]J. R. Kelly Jr., "A Taxonomic Survey of the Fishes of the Delta National Wildlife Refuge, with Emphasis on Distribution and Abundance," M.S. thesis, Louisiana State University, 1965.

[d]V. A. Guillory, "Distribution and Abundances of Fish in Thompson Creek and the Lower Mississippi River, Louisiana," M.S. thesis, Louisiana State University, 1974.

The third and final group of migratory species in the river—"euryhaline marine fishes" (46 species)—comprises species with wide range of salt tolerance (often from fresh- to full-complement seawater). These fishes typically invade the river from the Gulf of Mexico. Most are only seasonal inhabitants of the river, likely entering at times of low river discharge when saltwater wedges up river from the Gulf of Mexico (summer and fall). Marine invaders penetrate upriver to varying degrees, most rarely reaching New Orleans. However, a few are more tolerant of freshwater and penetrate much farther inland. The bull shark, a maneater, has been taken as far upriver as Alton, Illinois.[18]

Among the migratory (secondary freshwater, diadromous, and euryhaline marine) fishes in the river are sharks, two species of stingrays, the alligator gar, two species from the tarpon family, two species of eels, four species of herrings, a species of anchovy, two species of saltwater catfishes, three species of saltwater silversides, three species of livebearers, seven species of killifishes, two species of temperate basses, two species of jacks, eight species of drums, two species of porgies, three species of mullets, three species of sleepers, seven species of gobies, a hake, a searobin, a snapper, two species of mojarra, a mackerel, four species of left-eye flounders, two species of soles, and a tonguefish.

FACTORS INFLUENCING FISH DIVERSITY AND INDUSTRIAL IMPACTS

The variety of fish found in the lower Mississippi River is really quite remarkable, considering the degree of environmental disturbance. How is it that so many species of fish are able to live in such a disturbed reach of the river? One reason has to do with the immense size of the river and the variety of habitats it continues to offer. Although a quantitative temporal analysis of fish abundance trends in the river is not possible, it is clear that losses of species from the fish fauna have been minimal. Another reason for the high fish diversity is the transitional nature of the fish community near the river's terminus with the Gulf of Mexico. In addition to the seventy-six species of primary freshwater fish that regularly inhabit the river, fifty-eight species of migratory marine fish invade the river from near and offshore marine habitats of the Gulf of Mexico. Most of these species inhabit the river only in summer and fall months when river stage is low. They apparently retreat to more saline waters in winter and spring when the flow of the river is high. Thus, the lower corridor of the Mississippi River functions as an ecotone or transition zone between freshwater and marine ecosystems.

An ecotone develops everywhere different environments meet. Biotic communities in these areas typically are more diverse than communities from either of the primary environments, because they include species from both of the primary environments. The lower industrial corridor of the Mississippi River happens to be the area where the freshwater fish community of the river meets the marine fish community of the Gulf of Mexico. The mixing zone occurs in what is actually a freshwater environment, because marine fishes are generally more tolerant of freshwater than freshwater fishes are of saltwater. The tendency for saltwater to "wedge" up the river at times of low river flow may also be a contributing factor. Because saltwater is denser than freshwater, the salt wedge forms as a separate layer beneath the fresh surface water, with little mixing. Although a natural phenomenon in coastal rivers, the extent of saltwater intrusion up the lower Mississippi River delta is clearly greater today than in the past. This may be related to the deep draft channel being maintained for navigation, and the fact that the river is now discharging into deep waters of the Gulf of Mexico, near the edge of the continental shelf.

How are contaminants in the industrial corridor of the river likely affecting fishes that live there? A number of factors likely interact to limit the toxic effects of industrial contaminants on fish in the river. First, the total volume of water in the river is so great that contaminants rarely reach environmental concentrations that are lethal to fish (acute toxicity), even with the high volume of toxic discharges the Mississippi River receives. At low concentrations, contaminants mainly have sublethal effects (chronic toxicity), causing disease or affecting growth and reproduction.

Another factor that influences the toxic effects of contaminants in the river is their availability for uptake by fishes and other aquatic organisms. Most of the contaminants found in the river resist dissolution in water or tend to bind preferentially with sediment particles suspended in water. Fish may take up contaminants in this phase during the process of gill ventilation; however, most sediment-bound contaminants settle out to the river bottom with the sediment particles they are attached to. Here the contaminants may be buried, and thus remain unavailable for uptake, or they may be consumed by bottom-dwelling invertebrates, eventually making their way into the diets of fish.

The five species of suckers found in the river, together with the common carp (a member of the minnow family), have distinctive suctorial mouths, which they use to suck invertebrates from bottom sediments. In the process, they consume sediment-bound contaminants. This is why there is so much concern about consumption of bottom-feeding fish from the river.[19] It is unclear whether contaminants in the river are affecting the health of bottom-feeders such as suckers and carp. However, there are concerns about how environmental disturbances in the river may be influencing ecological interactions among these fishes. The carp is an Asian species that was introduced into the United States around the turn of the century. Because of its higher tolerance of environmental disturbance, it now outranks all of the native suckers in biomass production in the river. Concern is increasing that common carp, and two other Asian carp that have been introduced into the river in more recent times, will eventually competitively eliminate native suckers.

Position in the food chain is another ecological factor that influences an organism's contamination risk. Organisms near the top of the food chain typically have higher contaminant loads than organisms at lower levels. This process, known as bioaccumulation or biomagnification, occurs because organisms at higher positions accumulate contaminants from the organisms they feed on. The accumulation potential is highest for the organisms at the very top of the chain. This is why top predators in the river such as bass and gar tend to have the highest contaminant concentrations.

The amount of contaminants a fish can take up depends not only on how or where it feeds, but also on how long it lives and how long it remains in association with the contaminants. Primary freshwater fish are generally at greater risk of contaminant uptake than marine invaders, because they inhabit the river year 'round, whereas marine invaders' time in the river tends to be limited. Most of the fish and invertebrates found in the lower Mississippi River are relatively small and have short life cycles (less than two years). These organisms are likely not exposed to contaminants long enough to accumulate them in high concentrations. On the other hand, long-lived, large-bodied fishes, such as sturgeon and paddlefish, have the potential to accumulate much higher body burdens of contaminants than small fishes.

Sturgeon and paddlefish are less common in the Mississippi River today than they were in the past, but it is unclear if their decline is related to contamination stress. The two species of sturgeon found in the river—pallid and shovelnose sturgeon—are smaller than other North American sturgeons. These so-called river sturgeons are not harvested commercially for their eggs, which in other sturgeons are used to make caviar. The paddlefish, a distant relative of sturgeons, reaches a large size, and, like large sturgeons, was traditionally fished as a source of caviar. Though more common in the Mississippi River than sturgeon, paddlefish numbers are also declining. Today, the pallid sturgeon and paddlefish are federally protected species.

Fish have natural mechanisms for dealing with low levels of contamination. In some instances, toxic substances can be transformed into chemical forms that can eliminated from the fish's body or stored in fatty tissues where they pose little threat. In these instances, the only physiological consequence of exposure is the energetic costs of the biotransformation and elimination process. However, these costs can add up over time and affect growth and sexual maturation. In other instances, the contaminants cannot be transformed, or the transformation products are more toxic than the original contaminants. In these instances, the toxic effects are immediate.

Contaminants that interfere with the cycle or output of reproduction (e.g., estrogen-mimicking contaminants) have the potential to impact fish populations severely. Recent studies have identified instances where male fish exposed to common constituents of industrial and municipal effluent (e.g., alkylphenols) express female reproductive traits.[20] The impact of this abnormal sexual expression on reproductive success is largely unknown.

When humans regularly consume fish from the river, they put themselves at even greater risk of contaminant uptake than fishes that live in the river, because of their larger size, longer life span, and position at the top of the aquatic food chain. The lower Mississippi River no longer supports an active commercial fishery, but people living along the lower river (especially the poor) continue to use the river as a source of fish. The marsh and coastal waters adjacent to the river support a vast and bountiful commercial and recreational fishery. Contaminants in the river are undoubtedly making their way into food webs of the Gulf of Mexico. Thus, although advisories have been issued restricting consumption of fish from the river, humans likely are still taking in contaminants originating from the lower Mississippi River.

FUTURE WORK

Although available information does not permit a comprehensive retro-spective analysis of fish changes in the lower Mississippi River in relation to environmental changes that have occurred in the last century, it does pro-vide a useful baseline for assessing changes in the future. Long-term moni-toring of the biota of the lower industrial corridor of the river is one of the goals of a permanent river-monitoring station being proposed for develop-ment on the Mississippi River at English Turn, site of the Tulane University F. Edward Hebert "Riverside" Research Center. Sampling of the biota at this site, and others for which fish data was amassed in the past, will enable us to assess (a) changes in fish diversity since the 1970s and 1980s, (b) the re-lationship of these changes to specific water-quality changes, (c) the current extent of contamination of the biota, (d) the relationship of contamination to several key indicators of organismal, population, and community health, and (e) the consequences of the changes for river ecology. The study can make use of preserved specimens from previous surveys of the river. The specimens can be used to assess incidences of cancer and precancerous tumors, skeletal deformities, trophic interactions (gut contents analysis), parasitology, and abnormalities in growth, primary and secondary sexual expression, body proportions and symmetry, and several other biomarkers of contaminant exposure. Data gathered from specimens archived in the museum can be compared to data from specimens collected in the course of contemporary monitoring efforts.

Through this research we can develop a clearer understanding of how events of the past have impacted the biota of the lower Mississippi River and the threats that current activities in the river pose to continued persist-ence of the biota. We hope that the effort will guide ecologically and envi-ronmentally sound decisions about use of the river in the future.

Notes

Chapter 1. Introduction

1. For examples of urban environmental history, see *Journal of Urban History* 20, no. 3 (1994), which was a special issue on "The Environment and the City," guest edited by Christine M. Rosner and Joel A. Tarr; Andrew Hurley, ed., *Common Fields: An Environmental History of St. Louis* (St. Louis: Missouri Historical Society Press, 1997); and *Historical Geography* 25 (1997), an issue guest edited by Christopher Boone on "City and the Environment."

2. Carl Sauer, "The Morphology of Landscape," *University of California Publications in Geography* 2, no. 2 (1925): 19–54.

3. William B. Meyer, *Human Impact on the Earth* (New York: Cambridge University Press, 1996), esp. chap. 1.

4. William Cronon, *Nature's Metropolis: Chicago and the Great West* (New York: W. W. Norton, 1991).

5. For a discussion of this term, see Piers Blaikie and Harold Brookfield, eds., *Land Degradation and Society* (New York: Methuen, 1987).

6. See Sam B. Hilliard, *Man and the Environment in the Lower Mississippi*, Geoscience and Man 19 (Baton Rouge: Louisiana State University, Geoscience Publications, 1978).

7. Fred B. Kniffen, "The Lower Mississippi Valley: European Settlement, Utilization, and Modification," in *Cultural Diffusion and Landscapes*, ed. H. Jesse Walker and Randall Detro, Geoscience and Man 27 (Baton Rouge: Louisiana State University, Geoscience Publications, 1990), 3–34.

8. Peirce F. Lewis, *New Orleans: The Making of an Urban Landscape* (Cambridge, Mass.: Ballinger, 1976), 17.

Part 1. Transformation before Urbanization

1. William Cronon, *Changes in the Land: Indians, Colonists, and the Ecology of New England* (New York: Hill and Wang, 1983), 13.

2. William M. Denevan, "The Pristine Myth: The Landscape of the Americas in 1492," *Annals of the Association of American Geographers* 82 (1992): 381.

3. Peirce F. Lewis, *New Orleans: The Making of an Urban Landscape* (Cambridge, Mass.: Ballinger, 1976), 17.

Chapter 2. Making the City Inevitable

1. Kent Redford, "The Ecologically Noble Savage," *Orion Nature Quarterly* 9, no. 3 (1990): 24–29; William L. Balée, "Historical Ecology: Premises and Postulates," in *Advances in Historical Ecology*, ed. William L. Balée (New York: Columbia University Press, 1998), 16.

2. Balée, "Historical Ecology," 14–19.

3. René Dubos, *Beast or Angel?* (New York: Scribner's, 1974).

4. See the discussions in William L. Balée, ed., *Advances in Historical Ecology* (New York: Columbia University Press, 1998).

5. William M. Denevan, "The Pristine Myth: The Landscape of the Americas in 1492," *Annals of the Association of American Geographers* 82 (1992): 369–85; Martyn J. Bowden, "The Invention of American Tradition," *Journal of Historical Geography* 18 (1992): 3–26; Doug MacCleery, "Understanding the Role the Human Dimension Has Played in Shaping America's Forest and Grassland Landscapes," *Eco-Watch* (Feb. 10, 1994); William H. McNeill, "Mythhistory, or Truth, Myth, History and Historians," *American Historical Review* 91 (1986): 1–10; Peter Heehs, "Myth, History, and Theory," *History and Theory* 33 (1994): 1–19; Alfred W. Crosby, *Ecological Imperialism: The Biological Expansion of Europe, 900–1900* (Cambridge: Cambridge University Press, 1986); and Kirkpatrick Sale, *The Conquest of Paradise* (New York: Knopf, 1990).

6. Bowden, "Invention of American Tradition," 20.

7. MacCleery, "Understanding the Role," 7.

8. Bowden, "Invention of American Tradition," 20.

9. Tristram R. Kidder, "The Rat That Ate Louisiana," in *Advances in Historical Ecology*, ed. Balée, 141–68.

10. Balée, "Historical Ecology," 14–19; Tristram R. Kidder and William L. Balée, "Epilogue," in *Advances in Historical Ecology*, ed. Balée, 405–10.

11. Peirce F. Lewis, *New Orleans: The Making of an Urban Landscape* (Cambridge, Mass.: Ballinger, 1976), 17.

12. Kidder, "The Rat That Ate Louisiana."

13. William T. Penfound and Edward S. Hathaway, "Plant Communities in the Marshlands of Southeastern Louisiana," *Ecological Monographs* 8 (1938): 1–56; and David A. White, Steven P. Darwin, and Leonard B. Thien, "Plants and Plant Communities of Jean Lafitte National Historical Park, Louisiana," *Tulane Studies in Zoology and Botany* 24 (1983): 100–129.

14. Sherwood M. Gagliano, "Geoarchaeology of the Northern Gulf Shore," in *Perspectives on Gulf Coast Prehistory*, ed. D. D. Davis (Gainesville: University Presses of Florida, 1984), 1–40; and Roger T. Saucier, *Geomorphology and Quaternary Geologic History of the Lower Mississippi Valley*, 2 vols. (Vicksburg, Miss.: Mississippi River Commission, U.S. Army Corps of Engineers, 1994).

15. William H. Conner, Charles Sasser, and Nancy Barker, "Floristics of the Barataria Basin Wetlands, Louisiana," *Castanea* 51 (1986): 111–28; Mary E. Dunn, "Coquille Flora (Louisiana): An Ethnobotanical Reconstruction," *Economic Botany* 37 (1983):

349–59; Lionel N. Eleuterius, "The Marshes of Mississippi," *Castanea* 37 (1972): 153–68; Lionel N. Eleuterius and Sidney McDaniel, "The Salt Marsh Flora of Mississippi," *Castanea* 43 (1978): 86–95; and Penfound and Hathaway, "Plant Communities."

16. David A. Dell and Robert H. Chabreck, *Levees and Spoil Deposits as Habitat for Wild Animals in the Louisiana Coastal Marshes*, Research Report 7 (Baton Rouge: Louisiana Agricultural Experiment Station, 1986); Lionel N. Eleuterius and Ervin G. Otvos, "Floristic and Geologic Aspects of Indian Middens in Salt Marshes of Hancock County, Mississippi," *Sida* 8, no. 1 (1979): 102–12; Kidder, "The Rat That Ate Louisiana"; and Robert J. Lemaire, "A Preliminary Annotated Checklist of the Vascular Plants of the Marshes and Included Higher Lands of St. Bernard Parish, Louisiana," *Proceedings of the Louisiana Academy of Sciences* 24 (1961): 56–70.

17. Claire A. Brown, "The Vegetation of the Indian Mounds, Middens, and Marshes in Plaquemines and St. Bernard Parishes," in *Lower Mississippi River Delta: Reports on the Geology of Plaquemines and St. Bernard Parishes*, ed. R. J. Russell, H. V. Howe, J. H. McGuirt, C. F. Hohm, W. Hadley Jr., F. B. Kniffen, and C. A. Brown, Geological Bulletin 8 (New Orleans: Department of Conservation, Louisiana Geological Survey, 1936), 423–40.

18. Ibid.

19. Gagliano, "Geoarchaeology of the Northern Gulf Shore"; William G. McIntire, *Prehistoric Indian Settlements of the Changing Mississippi River Delta*, Coastal Studies Series 1 (Baton Rouge: Louisiana State University Press, 1958).

20. Brown, "Vegetation of the Indian Mounds"; Dunn, "Coquille Flora"; Eleuterius and Otvos, "Floristic and Geologic Aspects"; Kidder, "The Rat That Ate Louisiana"; and Lemaire, "Preliminary Annotated Checklist."

21. Gagliano, "Geoarchaeology of the Northern Gulf Shore"; and Christopher Neill and Linda A. Deegan, "The Effects of Mississippi River Lobe Development on the Habitat Composition and Diversity of Louisiana Coastal Wetlands," *American Midland Naturalist* 116 (1986): 296–303.

22. Tristram R. Kidder, "Perspectives on the Geoarchaeology of the Lower Mississippi Valley," *Engineering Geology* 45 (1996): 305–23.

23. Dave D. Davis, "Comparative Aspects of Late Prehistoric Faunal Ecology at the Sims Site," *Louisiana Archaeology* 11 (1987): 111–38; Kenneth R. Jones, Herschel A. Franks, and Tristram R. Kidder, *Cultural Resources Survey and Testing for Davis Pond Freshwater Diversion, St. Charles Parish, Louisiana*, 2 vols. Report No. COELMN/PD-93/01, Cultural Resources Series (New Orleans: Earth Search, 1994); and Tristram R. Kidder, *Archaeological Data Recovery at 16JE218, Jefferson Parish, Louisiana*, Report No. COELMN/PD-95/03, Cultural Resources Series (New Orleans: Earth Search, 1995).

24. Sherwood M. Gagliano, Susan Fulgham, and Bert Rader, *Cultural Resources Studies in the Pearl River Mouth Area, Louisiana-Mississippi* (Baton Rouge: Coastal Environments, 1979), 2.19–22; and Saucier, *Geomorphology*; J. Richard Shenkel, "Big Oak and Little Oak Islands: Excavations Interpretations," *Louisiana Archaeology* 1 (1974): 37–65; Shenkel, *Oak Island Archaeology: Prehistoric Estuarine Adaptations in the Mississippi River Delta* (New Orleans: National Park Service, Jean Lafitte National Historical Park,

1981); Shenkel, "Early Woodland in Coastal Louisiana," in *Perspectives on Gulf Coast Prehistory*, ed. Davis, 41–71; and J. Richard Shenkel and Jon L. Gibson, "Big Oak Island: An Historical Perspective of Changing Site Function," *Louisiana Studies* 8 (1974): 173–86.

25. J. Richard Czajkowski, "Preliminary Report of Archeological Investigations in Orleans Parish," *Louisiana Conservation Review* 4 (1934): 12–18; Gagliano, Fulgham, and Rader, *Cultural Resources Studies*, 2.19–22; Shenkel, *Oak Island Archaeology*; and Shenkel, "Early Woodland in Coastal Louisiana."

26. Shenkel, *Oak Island Archaeology*, 7–9, and plate 2.

27. Gagliano, Fulgham, and Rader, *Cultural Resources Studies*, Table 2.2.

28. Shenkel, *Oak Island Archaeology*.

29. Lewis, *New Orleans*.

30. Betsy Swanson, *Terre Haute De Barataria*, Monograph 11 (Harahan, La.: Jefferson Parish Historical Commission, 1991); and Carlos Trudeau, "Plan del Local de las Tierras que Rodean la Ciudad de Nueva Orleans," Historic New Orleans Collection, New Orleans, 1803.

31. Czajkowski, "Preliminary Report"; James A. Ford and George I. Quimby Jr., "The Tchefuncte Culture: An Early Occupation of the Lower Mississippi Valley," *Memoirs of the Society for American Archaeology* 2 (1945); Kidder, "The Rat That Ate Louisiana"; Shenkel, "Big Oak and Little Oak Islands"; Shenkel, *Oak Island Archaeology*; Shenkel, "Early Woodland in Coastal Louisiana"; and Shenkel and Gibson, "Big Oak Island."

32. George B. Davis, Leslie J. Perry, and Joseph Kirkley, *Atlas to Accompany the Official Records of the Union and Confederate Armies* (Washington, D.C.: GPO, 1891–95), plate XCI; and John W. Foster, *Pre-Historic Races of the United States of America* (Chicago: S. C. Griggs, 1874), 157–58.

33. Swanson, *Terre Haute De Barataria*, 17; and Trudeau, "Plan del Local de las Tierras."

34. Roger T. Saucier, *Recent Geomorphic History of the Pontchartrain Basin*, Coastal Study Series 9 (Baton Rouge: Louisiana State University, 1963); and Roger T. Saucier, "Archaeological Significance of Large Ephemeral Lakes and Lake Outlets of the Mississippi River Deltaic Plain," *Louisiana Archaeology* 23 (1999): 1–36. For a complete discussion see Saucier, *Geomorphology*; Saucier, *Recent Geomorphic History*; and Shenkel, *Oak Island Archaeology*.

35. Saucier, *Recent Geomorphic History*.

36. See Lewis, *New Orleans*, fig. 7.

37. Saucier, *Recent Geomorphic History*, figs. 31–32.

38. Edna B. Freiberg, *Bayou St. John in Colonial Louisiana, 1699–1803* (New Orleans: Harvey Press, 1980); Jay Higginbotham, trans. and ed., *The Journal of Sauvole* (Mobile, Ala.: Colonial Books, 1969), 30, 34; Pierre Le Moyne d'Iberville, *Iberville's Gulf Journal*, trans. and ed. Richebourg G. McWilliams (University: University of Alabama Press, 1981), 57; and Pierre Margry, *Découvertes et Établissements des Français dans L'Ouest et dans le Sud de L'Amérique Septentrionale (1694–1703)*, Tome 4 (Paris: Maisonneuve, 1881), 165–66.

39. Marco J. Giardino, "Documentary Evidence for the Location of Historic Indian Villages in the Mississippi Delta," in *Perspectives on Gulf Coast Prehistory*, ed. Davis,

244; Bernard de La Harpe, *Journal Historique de l'Establisement des Francais a la Louisiane* (New Orleans: A. L. Boimare, 1831), 21, 27; Baron Marc de Villiers, "A History of the Foundation of New Orleans (1717–1720)," trans. Warrington Dawson, *Louisiana Historical Quarterly* 3 (April 1920): 183; La Harpe, *Journal Historique*, 21; and T. Jeffreys, "The Course of Mississippi River from Bayagoulas to the Sea," 1759, Louisiana Collection, Howard Tilton Memorial Library, Tulane University, New Orleans.

40. Trudeau, "Plan del Local de las Tierras."

41. Davis, Perry, and Kirkley, *Atlas*, plate XCI; Saucier, *Recent Geomorphic History*, figs. 30 and 32.

42. Villiers, "History of the Foundation," 166; Giardino, "Documentary Evidence," 236.

43. Giardino, "Documentary Evidence"; Higginbotham, trans. and ed., *Journal of Sauvole*; Richebourg G. McWilliams, trans. and ed., *Fleur de Lys and Calumet: Being the Penicaut Narrative of French Adventure in Louisiana* (Baton Rouge: Louisiana State University Press, 1953); d'Iberville, *Iberville's Gulf Journal*, 111; Margry, *Découvertes et Établissements*, 399, 455; and John R. Swanton, *Indian Tribes of the Lower Mississippi Valley and Adjacent Coast of the Gulf of Mexico*, Bulletin 43 (Washington, D.C.: Bureau of American Ethnology, 1911).

44. Maurice Ries, "The Mississippi Fort, Called Fort de la Boulaye," *Louisiana Historical Quarterly* 19 (1936): map 3.

45. Villiers, "History of the Foundation," 164; see also Giardino, "Documentary Evidence," 248–49.

46. Antoine Le Page Du Pratz, *Histoire de la Louisiane*, 3 vols. (Paris: De Bure, La Veuve Delaguette, Lambert, 1758), 46, 82.

47. Giardino, "Documentary Evidence," 243–44.

48. Freiberg, *Bayou St. John*; and Christopher Matthews, personal communication with author, 1999.

49. Giardino, "Documentary Evidence," 243–44; and Jean-Baptiste LeMoyne Bienville, "Sauvages de la Louisianne leur Nombre, et le Commerce Qu'on pour Faire Avec Eux," n.d., Ayer Collection, Newberry Library, Chicago.

50. Kidder, *Archaeological Data Recovery at 16JE218, Jefferson Parish, Louisiana.*

Chapter 3. Impenetrable but Easy

1. Pierre Le Moyne d'Iberville, *Iberville's Gulf Journals*, trans. and ed. Richebourg Gaillard McWilliams (University: University of Alabama Press, 1981), 56.

2. Lauren C. Post, "The Domestic Animals and Plants of French Louisiana as Mentioned in the Literature with Reference to Sources, Varieties and Uses," *Louisiana Historical Quarterly* 16 (Oct. 1933): 559–60. *Mississippi Provincial Archives: French Dominion*, 5 vols.: vols. 1–3, ed. Dunbar Rowland and Albert Godfrey Sanders (Jackson: Press of the Mississippi Department of Archives and History, 1929–32); vols. 4–5, ed. Dunbar Rowland, Albert Godfrey Sanders, and Patricia Kay Galloway (Baton Rouge: Louisiana State University Press, 1984), 2:10 (hereafter *MPAFD*).

3. In March 1699 Iberville encountered a herd of over 200 buffalo grazing along

Lake Maurepas. He later found herds at Baie St. Louis and at Lake Pontchartrain. Iberville, *Iberville's Gulf Journals,* 82, 112, 113, 142.

4. McWilliams, trans. and ed., *Fleur de Lys and Calumet,* 20.

5. Ibid., 254–55.

6. Recently, William Cronon has made the argument, most forcefully and eloquently, that Indians altered the natural environment in many ways, with lasting impact for European settlers. William Cronon, *Changes in the Land: Indians, Colonists, and the Ecology of New England* (New York: Hill and Wang, 1983).

7. Daniel H. Usner Jr., *Indians, Settlers, and Slaves in a Frontier Exchange Economy: The Lower Mississippi Valley before 1783* (Chapel Hill: University of North Carolina Press, for the Institute of Early American History and Culture, 1992), 21–24; Usner, *American Indians in the Lower Mississippi Valley: Social and Economic Histories* (Lincoln: University of Nebraska Press, 1998), 33–55; Jeffrey P. Brain, *On the Tunica Trail,* 3rd ed., Anthropological Study Series 1 (Baton Rouge: Department of Culture, Recreation and Tourism, Louisiana Archaeological Survey and Antiquities Commission, 1994); Brain, *Tunica Archaeology,* Papers of the Peabody Museum of Archaeology and Ethnology, Harvard University, no. 78 (Cambridge: Harvard University Press, 1988); Brain, "Late Prehistoric Settlement Patterning in the Yazoo Basin and Natchez Bluffs Regions of the Lower Mississippi Valley," in *Mississippian Settlement Patterns,* ed. Bruce D. Smith (New York: Academic Press, 1978), 331–68.

8. Robert to Pontchartrain, Nov. 1708, *MPAFD,* 2:45–46.

9. [Eléonore Oglethorpe, Marquise de Mézières,] Paris, [to Jean Gravé de La Mancelière, en route to Louisiana], June 12, 1721, Rosemonde E. and Emile Kuntz Collection, Tulane University Special Collections, Jones Hall, Tulane University, New Orleans.

10. La Mothe Cadillac to Pontchartrain, Oct. 1713, *MPAFD,* 2:166–70, 198.

11. On continuities between Indian and European settlement patterns, see John Mack Faragher, *Sugar Creek: Life on the Illinois Prairie* (New Haven: Yale University Press, 1986), 3–9.

12. Duclos to Pontchartrain, Oct. 1713, *MPAFD,* 2:80, 2:208.

13. Bienville to Pontchartrain, Feb. 25, 1708, *MPAFD,* 3:122.

14. D'Artaguette to Pontchartrain, Feb. 1710, *MPAFD,* 2:52–53.

15. Jay K. Ditchy, "Early Census Tables of Louisiana," *Louisiana Historical Quarterly* 13 (April 1930): 209, 218–20, 227; Duclos to Pontchartrain, Oct. 1713, *MPAFD,* 2:80–81; Heloise H. Crozat, trans., "Louisiana in 1724: Banet's Report to the Company of the Indies, Dated Paris, Dec. 20, 1724," *Louisiana Historical Quarterly* 12 (Jan. 1929): 122–26; Post, "Domestic Animals and Plants of French Louisiana," 563–64.

16. Sam B. Hilliard conservatively estimates a 20 percent yield from cattle herds. See Hilliard, *Hog Meat and Hoecake: Food Supply in the Old South* (Carbondale: Southern Illinois University Press, 1972), 128–29.

17. De La Chaise to the Directors of the Company of the Indies, Sept. 1723, *MPAFD,* 2:312.

18. As early as 1709 French officials noted the suitability of lower Louisiana for

rice agriculture, but they would wait seven more years before introducing the crop to the colony. Pontchartrain to d'Artaguette, July 1709, *MPAFD*, 2:130–31; D'Artaguette to Pontchartrain, Feb. 1710, *MPAFD*, 2:59; D'Artaguette to Pontchartrain, May 1712, *MPAFD*, 2:63.

19. Marcel Giraud, *A History of French Louisiana*, 5 vols., trans. Brian Pearce (Baton Rouge: Louisiana State University, 1993), 2:134–36; D'Artaguette to Pontchartrain, Feb. 1710, *MPAFD*, 2:52–53; D'Artaguette to Pontchartrain, May 1712, *MPAFD*, 2:61; Ditchy, "Early Census Tables of Louisiana," 214–29; Bienville and Salmon to Maurepas, May 1733, *MPAFD*, 3:600.

20. Post, "Domestic Animals and Plants of French Louisiana," 575–76; Lewis Cecil Gray, *History of Agriculture in the Southern United States to 1860*, 2 vols. (Gloucester, Mass.: Peter Smith, 1958), 66; Sam B. Hilliard, "Antebellum Tidewater Rice Culture in South Carolina and Georgia," in *European Settlement and Development in North America: Essays on Geographical Change in Honour and Memory of Andrew Hill Clark*, ed. James R. Gibson (Toronto: University of Toronto Press, 1978), 94, 110.

21. Land grant, Company of the Indies to Bienville, March 27, 1719, Rosemonde E. and Emile Kuntz Collection, Tulane Special Collections, Tulane University, New Orleans.

22. Ditchy, "Early Census Tables of Louisiana," 222–23. In the original document Kolly's rice production is measured in *quartes*. A *quarte* was a French measurement of volume commonly used for grain, and was roughly equal to one U.S. bushel. Ronald Edward Zupko, *French Weights and Measures before the Revolution* (Bloomington: Indiana University Press, 1978), 147–48; Post, "Domestic Animals and Plants of French Louisiana," 576.

23. Hubert to the Council [1717], *MPAFD*, 2:229, 233.

24. Minutes of the Council of Commerce, July 1721, *MPAFD*, 2:266; Baron Marc de Villiers, "A History of the Foundation of New Orleans (1717–1722)," trans. Warrington Dawson, *Louisiana Historical Quarterly* 3 (April 1920): 195. It should be noted that in 1718 the first experiments at planting tobacco at Natchez commenced. See Giraud, *History of French Louisiana*, 5:133. Tobacco planting at Natchez further suggested the need for a port on the Mississippi River, though not necessarily one so far south as New Orleans. It was the success of rice planting that demonstrated the need for a port in the lower reaches of the river.

25. Villiers, "History of the Foundation."

26. Alfred Crosby, *Ecological Imperialism: The Biological Expansion of Europe, 900–1900* (Cambridge: Cambridge University Press, 1986), 146–70, 179–80; Christopher Morris, "How to Prepare Buffalo and Other Things French Taught Indians about Nature," in *Colonial Louisiana: A Tricentennial Symposium,* ed. Bradley G. Bond (Baton Rouge: Louisiana State University Press, forthcoming).

27. Gustavus Schmidt, trans., "O'Reilly's Ordinance of 1770," *Louisiana Historical Quarterly* 11 (April 1928): 239; "Governor Carondelet's Levee Ordinance of 1792," trans. Laura L. Porteous, *Louisiana Historical Quarterly* 10 (Oct. 1927): 514.

28. Post, "Domestic Animals and Plants of French Louisiana," 564.

29. Minutes of the Council, January 1723, *MPAFD*, 2:282, 285–86.

30. On agriculture and capitalism in early America, see Michael Merrill, "Cash Is Good to Eat: Self-Sufficiency and Exchange in the Rural Economy of the United States," *Radical History Review*, no. 3 (1977): 42–71; James A. Henretta, "Families and Farms: *Mentalité* in Pre-Industrial America," *William and Mary Quarterly*, 3d ser., 35 (1978): 3–32; Christopher Clark, *Roots of Rural Capitalism: Western Massachusetts, 1780–1860* (Ithaca, N.Y.: Cornell University Press, 1990); Winifred B. Rothenberg, *From Market-places to a Market Economy: The Transformation of Rural Massachusetts, 1750–1850* (Chicago: University of Chicago Press, 1992). On the slow development of capitalism in France and its consequences for North American colonization, see Denys Delâge, *Amerindians and Europeans in Northeastern North America, 1600–64*, trans. Jane Brierley (Vancouver: University of British Columbia Press, 1993), 3–35. On the exchange economy in which many settlers participated within colonial Louisiana, and which delayed the development of plantation agriculture, see Usner, *Indians, Settlers, and Slaves in a Frontier Exchange Economy.*

31. Gwendolyn Midlo Hall, *Africans in Colonial Louisiana: The Development of Afro-Creole Culture in the Eighteenth Century* (Baton Rouge: Louisiana State University Press, 1992), 60, 175; Post, "Domestic Animals and Plants of French Louisiana," 576.

32. In 1721 thirteen slave owners owned at least ten slaves. Nine of them also owned at least eight cattle. The best example of the connection between cattle and slaves was at Tchopitoulas, where two farmers each owned thirty-three slaves and between them owned ninety cattle. The census also shows that Jean-Daniel Kolly owned forty-six slaves but just five cattle. Kolly may have been an exception; however, it may also be that the census was mistaken. At his death in 1729 Kolly owned sixty-five slaves and forty cattle. See Ditchy, "Early Census Tables of Louisiana," 214–20; Giraud, *History of French Louisiana*, 5:164. Usner, *Indians, Settlers, and Slaves in a Frontier Exchange Economy*, 182–83. On the connection between cattle and slave ownership in another setting, see Christopher Morris, *Becoming Southern: The Evolution of a Way of Life, Vicksburg and Warren County, Mississippi, 1770–1860* (New York: Oxford University Press, 1995), 23–41, 189–91.

33. In 1727 and 1743 the colonial government issued official edicts requiring the construction of levees. Giraud, *History of French Louisiana*, 5:193, 194, 206, 208–9; Robert W. Harrison, *Alluvial Empire: A Study of State and Local Efforts Toward Land Development in the Alluvial Valley of the Lower Mississippi River, Including Flood Control, Land Drainage, Land Clearing, Land Forming*, 2 vols. (Little Rock, Ark.: Pioneer Press, for the Economic Research Service, U.S. Department of Agriculture, 1961), 1:53–54; Périer and de la Chaise to the Directors of the Company of the Indies, Nov. 1727, *MPAFD*, 2:547, 565; Etienne de Périer and de la Chaise to the Directors of the Company of the Indies, Aug. 1728, *MPAFD*, 2:589–90.

34. Giraud, *History of French Louisiana*, 5:192; Gray, *History of Agriculture in the Southern United States to 1860*, 2:66. Throughout much of the eighteenth century South Carolina rice planters endeavored to control river and swamp waters in a similar man-

ner. By the end of the century, however, they harnessed the power of coastal tides to control irrigation of fields, a technique not used in Louisiana. See Joyce E. Chaplin, *An Anxious Pursuit: Agricultural Innovation and Modernity in the Lower South, 1730–1815* (Chapel Hill: University of North Carolina Press, for the Institute of Early American History and Culture, 1993), 228–32.

35. Giraud, *History of French Louisiana*, 5:195.

36. Noting that settlers needed to keep slaves occupied year round, the Council in New Orleans encouraged company directors to facilitate a timber trade with St. Domingue. See Council to the Directors, Aug. 1725, *MPAFD*, 2:494; John Hebron Moore, "The Cypress Lumber Industry of the Lower Mississippi Valley During the Colonial Period," *Louisiana History* 24 (1983): 25–47; Giraud, *History of French Louisiana*, 5:142–43; Périer to the Abbé Raguet, Aug. 1728, *MPAFD*, 2:616; Hilliard, "Antebellum Tidewater Rice Culture in South Carolina and Georgia," 98.

37. Moore, "Cypress Lumber Industry of the Lower Mississippi Valley," 590; Giraud, *History of French Louisiana*, 5:193–94.

38. Périer to Maurepas, Aug. 1730, *MPAFD*, 4:38.

39. Bienville and Salmon to Maurepas, March 1734, *MPAFD*, 3:637–38.

40. Bienville to the Council, Feb. 1723, *MPAFD*, 3:343.

41. Usner, *Indians, Settlers, and Slaves in a Frontier Exchange Economy*, 198–201.

42. Giraud, *History of French Louisiana*, 5:213–16. Usner, *Indians, Settlers, and Slaves in a Frontier Exchange Economy*, 40–41, 115.

43. Villiers, "History of the Foundation," 186, 190.

Part 2. Environment in Service of the City

1. William Cronon, *Nature's Metropolis: Chicago and the Great West* (New York: W. W. Norton, 1991), 26–47.

2. A full discussion of the levee building appears in Albert E. Cowdrey, *Land's End: A History of the New Orleans District, U.S. Army Corps of Engineers* (New Orleans: U.S. Army Corps of Engineers, 1977).

Chapter 4. Forests and Other River Perils

1. *Louisiana Gazette*, March 26, 1817.

2. See *Heirs of Fulton and Livingston v. Henry M. Shreve*, United States District Court for New Orleans, in Federal Records Center in Forth Worth, Texas, accession number 70A-006, file number 1003, box number 2188.

3. Shreve's log from that journey upriver is reprinted in the *Louisiana Gazette* of May 6, 1817.

4. Morris Birkbeck, *Notes on a Journey in America* (London: Severn and Co., 1818), 85.

5. There are two major biographies on Henry Shreve. Both are engaging but

dated: Florence Dorsey, *Master of the Mississippi* (Boston: Houghton-Mifflin, 1941); and Edith McCall, *Conquering the Rivers* (Baton Rouge: Louisiana State University Press, 1984).

6. Francois Xavier Charlevoix, *Journal of a Voyage to North America,* vol. 2, ed. and trans. Louise Phelps Kellog (Chicago: The Caxton Club, 1923), 258.

7. Almost every author that wrote about New Orleans during the eighteenth and nineteenth centuries gushed about its natural advantages for commerce. A short sampling of these sources includes Thomas Ashe, *Travels in America,* vol. 3 (London: Richard Phillips, 1808); Baird, *View of the Valley of the Mississippi;* Jonathan Carver, *Travels Through the Interior Parts of North America, in the Years 1766, 1767, and 1768* (Minneapolis: Ross and Haines, 1956); Charlevoix, *Journal of a Voyage to North America,* vol. 2; Andrew Ellicott, *The Journal of Andrew Ellicott* (Philadelphia: Budd and Hartram, 1803); Thomas Hutchins, *An Historical Narrative and Topographical Description of Louisiana and West Florida* (Philadelphia: Robert Aitken, 1784); Henry A. Murray, *Lands of the Slave and the Free* (London: John W. Parker and Son, 1855); James Pitot, *Observations on the Colony of Louisiana from 1796 to 1802* (Baton Rouge: Louisiana State University Press, 1979); C. C. Robin, *Voyage to Louisiana, 1803–1805,* trans. Stuart O. Landry Jr. (New Orleans: Pelican Publishing, 1966); Christian Schultz, *Travels on an Inland Voyage,* vol. 2 (New York: Isaac Riley, 1810); Charles Sealsfield, *The Americans as They Are* (London: Hurst, Chance, and Co., 1828); James Stuart, *Three Years in North America* (New York: J. and J. Harper, 1833); and Lady Emmeline Stuart Wortley, *Travels in the United States,* vol. 1 (London: Richard Bentley, 1851).

8. George Rogers Taylor, *The Transportation Revolution, 1815–1860* (White Plains, N.Y.: M. E. Sharpe, 1951), 143.

9. For a dated but useful source on river traffic prior to the era of steam, see Leland D. Baldwin, *The Keelboat Age on Western Waters* (Pittsburgh: University of Pittsburgh Press, 1941).

10. Mechanical innovators had already spent centuries tinkering with steam engines. For information on the invention of the steam engine, see R. A. Buchanan, *The Industrial Archaeology of the Stationary Steam Engine* (London: Allen Lane, 1976); H. W. Dickinson, *A Short History of the Steam Engine* (Cambridge: Cambridge University Press, 1938); W. H. Fowler, *Stationary Steam Engines* (Manchester: Scientific Publishing, 1908); Richard L. Hills, *Power from Steam* (Cambridge: Cambridge University Press, 1989); J. P. Muirhead, *The Origins and Progress of the Mechanical Inventions of James Watt* (London: J. Murray, 1854); A. J. Pacey, *The Maze of Ingenuity* (London: Allen Lane, 1974); R. H. Thurston, *A History of the Growth of the Steam Engine* (New York: D. Appleton, 1878).

11. For the text of this monopoly, see William Claiborne to Robert Livingston, March 26, 1811, in *W.C.C. Claiborne, Official Letter Books,* ed. Dunbar Rowland (Jackson, Miss.: State Department of Archives and History, 1917), 5:192; chapter 26, *Acts of the Second Session, Third Legislature of the Territory of Orleans,* April 19, 1811, 112. For the impact of this law on trade in the West, see H. Dora Stecker, "Constructing a Navigation System in the West," *Ohio Archeological and Historical Quarterly* 22 (1913): 18–22.

12. For information on the first voyage of the *New Orleans*, see E. W. Gould, *Fifty Years on the Mississippi, or, Gould's History of River Navigation* (Saint Louis: Nixon-Jones Printing, 1889), 82–84; Talbot Hamlin, *Benjamin Henry Latrobe* (New York: Oxford University Press, 1955), 370–73; J. H. B. Latrobe, "The First Steamboat Voyage on the Western Waters," in the Mississippi River Papers, Howard-Tilton Memorial Library Special Collections, Manuscript Division, Tulane University; McCall, *Conquering the Rivers*, 83–85.

13. For a detailed examination of the development of the western steamboat, see Louis Hunter, "The Invention of the Western Steamboat," *Journal of Economic History* 3 (May 1943).

14. For an examination of why the Fulton group's vessels traveled such tightly circumscribed routes, see what remains the finest book available on steamboats: Louis Hunter, *Steamboats on the Western Rivers* (Cambridge: Harvard University Press, 1949). Many early nineteenth-century historians of steam argued that the *New Orleans* could not have made the journey upriver from New Orleans as far as Louisville, but Hunter states that economics motivated the Mississippi Steamboat Navigation Company to limit their route to the New Orleans–Natchez trade. Support for that point of view comes from news of the *New Orleans*'s arrival in the Crescent City. The *Louisiana Gazette* of January 13, 1812, noted that the steamboat was "intended as a regular trader between this [New Orleans] and Natchez." It would seem from that comment that the Fulton group never had any intention of taking their boat further upstream than Natchez, and thus the short route was not the result of the boat's limited capabilities.

15. James Hall, *Statistics of the West, at the Close of the Year 1836* (Cincinnati: J. A. James and Co., 1836), 232.

16. Between 1812 and 1817 steam technology had not languished in the valley despite the Fulton group's monopoly. The monopolists themselves had brought vessels other than the *New Orleans* to the Mississippi. See, for instance, news of the *Vesuvius*: *Niles' Weekly Register* 5 (Dec. 18, 1813), 272. See news of the Fulton group's vessel the *Aetna*: *Lexington Reporter*, Jan. 31, 1816. Shreve also had brought steamboats to the river system prior to the *Washington*. For instance, on the voyage of the *Enterprise*, see *Niles' Weekly Register* 8 (July 1, 1815): 320.

17. *Heirs of Fulton and Livingston v. Henry M. Shreve*, United States District Court for New Orleans, in Federal Records Center in Forth Worth, Texas, accession number 70A-006, file number 1003, box number 2188.

18. Records of the New Orleans Collector of Levee Dues, Register of flatboats, barges, rafts, and steamboats in the port of New Orleans, 1806–1823, in City Archives, Louisiana Division, New Orleans Public Library.

19. Records of the New Orleans Collector of Levee Dues, Register of flatboats, barges, rafts, and steamboats in the port of New Orleans, 1806–1823, in City Archives, Louisiana Division, New Orleans Public Library; *House Ex. Doc. 6*, Part 2, 50th Cong., 1st sess., 199.

20. *Niles' Weekly Register* 33 (Nov. 17, 1827), 181; *House Ex. Doc. 6*, Part 2, 50th Cong., 1st sess., 199.

21. *Senate Ex. Doc. 42*, 32nd Cong., 1st sess., 15; *House Ex. Doc. 6*, Part 2, 50th Cong., 1st sess., 213, 221.

22. See William Cronon, *Nature's Metropolis* (New York: W. W. Norton, 1991).

23. *House Ex. Doc. 6*, Part 2, 50th Cong., 1st sess., 191, 199, 201–2, 215; Lester B. Shippe, *Bishop Whipple's Southern Diary, 1843, 1844* (Minneapolis: University of Minnesota Press, 1937), 95.

24. Charles Murray, *Travels in North America* (London: Richard Bentley, 1839), 2:189.

25. Shippe, *Bishop Whipple's Southern Diary*, 95.

26. Timothy Flint, *A Condensed Geography and History of the Western States* (Cincinnati: E. H. Flint, 1828), 238.

27. Ralph K. Andrist, *Steamboats on the Mississippi* (New York: American Heritage Publishing, 1962), 115.

28. Dorsey, *Master of the Mississippi*, 129; McCall, *Conquering the Rivers*, 154; Hunter, *Steamboats on the Western Rivers*, 19.

29. For the record times, see James T. Lloyd, *Lloyd's Steamboat Directory, and Disasters on the Western Waters* (Philadelphia: J. T. Lloyd, 1856), 292–93.

30. Flint, *Condensed Geography and History of the Western States*, 46.

31. Baird, *View of the Valley of the Mississippi*, 280, 340.

32. Timothy Flint, "A Condensed Geography and History of the Western States," in *Before Mark Twain*, ed. John Francis McDermott (Carbondale: Southern Illinois University Press, 1968), 10.

33. Henry Tudor, *Narrative of a Tour of North America* (London: James Duncan, 1834), 2:36.

34. For several exponents of this view, see John Kasson, *Civilizing the Machine* (New York: Viking Press, 1976); Leo Marx, *The Machine in the Garden* (New York: Oxford University Press, 1964); David E. Nye, *American Technological Sublime* (Cambridge, Mass.: MIT Press, 1994); Wolfgang Schivelbusch, *The Railway Journey* (Berkeley: University of California Press, 1977); and Richard White, *The Organic Machine* (New York: Hill and Wang, 1995).

35. *New Orleans Bee*, Feb. 14, 1823.

36. Lloyd, *Lloyd's Steamboat Directory*, 61.

37. Ibid., 61, 63

38. *House Ex. Doc. 35*, 17th Cong., 2nd sess., 21.

39. Wortley, *Travels in the United States*, 1:215.

40. John Bradbury, "Travels in the Interior of America, in the Years 1809, 1810, and 1811," in *Early Western Travels*, ed. Reuban Gold Thwaites (Cleveland: Arthur H. Clark, 1904), 200.

41. T. L. Nichols, *Forty Years of American Life* (London: Longmans, Green, & Co., 1874), 122.

42. *Niles' Weekly Register* 38 (March 27, 1830): 97.

43. Hunter, *Steamboats on the Western Rivers*, 193.

44. *House Doc. 379*, 21st Cong., 1st sess., 1

45. *Senate Ex. Doc. 2*, 19th Cong., 1st sess., 52.

46. Hunter, *Steamboats on the Western Rivers*, 194.

47. Dorsey, *Master of the Mississippi*, 15.

48. *Niles' Weekly Register* 70 (April 11, 1846): 132.

49. *House Ex. Doc. 6*, Part 2, 50th Cong., 1st sess., 198.

50. *New Orleans Bee*, May 10, 1836.

51. *House Doc. 11*, 20th Cong., 1st sess., 4; *Senate Ex. Doc. 1*, 23rd Cong., 1st sess., 128; *Senate Ex. Doc. 1*, 23rd Cong., 2nd sess., 163; *Senate Ex. Doc. 1*, 24th Cong., 1st sess., 168; *Senate Doc. 1*, 29th Cong., 1st sess., 347.

52. James Hall, *Notes on the Western States* (Philadelphia: Harrison Hall, 1838), 238.

53. *Senate Ex. Doc. 42*, 32nd Cong., 1st sess., 113.

54. F. Terry Norris, "Where Did the Villages Go?," in *Common Fields: An Environmental History of St. Louis*, ed. Andrew Hurley (St. Louis: Missouri Historical Society Press, 1997), 81.

55. *Senate Ex. Doc. 42*, 32nd Cong., 1st sess., 15; *House Ex. Doc. 6*, Part 2, 50th Cong., 1st sess., 213, 221.

56. *House Rep. 75*, 18th Cong., 1st sess., 3.

57. *House Ex. Doc. 66*, 23rd Cong., 2nd sess., 3.

58. Charles Lyell, *A Second Visit to the United States of North America* (London: John Murray, 1849), 1:344. In the second volume of his travel narrative, Lyell criticized Shreve's plan for cutting the trees, noting that: "The roots also of trees growing at the edge of the stream, were very effective formerly in holding the soil together, before so much timber had been cleared away. Now the banks offer less resistance to the wasting action of the stream" (2:171).

59. *Acts Passed at the Second Session of the Fifteenth Legislature of the State of Louisiana* (New Orleans: J. C. De St. Romes, 1842), 280.

Chapter 5. Subduing Nature through Engineering

1. Paul Octave Hébert, *Annual Report of P. O. Hébert, State Engineer to the Legislature*, Jan. 1847 (New Orleans, 1847), 8–14; Absalom D. Wooldridge, "Report of the Internal Improvements of Louisiana," Appendix, Jan. 1850, in *Documents of the First Session of the Third Legislature of the State of Louisiana* (copy in Hill Memorial Library, Louisiana State University, Baton Rouge), 8–16. For further discussion of the debate among engineers who supported a system of outlets, see George S. Pabis, "Delaying the Deluge: The Engineering Debate over Flood Control on the Lower Mississippi River, 1846–1861," *Journal of Southern History* 64 (Aug. 1998): 425–30.

2. George Willard Reed Bayley, "Overflow of the Delta of the Mississippi: Review of Charles Ellet's Report on the Overflows of the Delta of the Mississippi," *De Bow's Review* 13 (Aug. 1852): 173.

3. Michael Thomas Meier, "Caleb Goldsmith Forshey: Engineer of the Old Southwest, 1813–1881," Ph.D. diss., Memphis State University, 1982, 5–16. I want to thank Michael Meier for several years of his friendship. I have based many of my ideas about the history of Mississippi River flood control on my discussions with him and his excellent scholarship.

4. *House Miscellaneous Documents,* 42nd Cong., 3rd sess., no. 41: Caleb G. Forshey, Memorial of Citizens of the State of Louisiana, in Favor of Nationalizing the Levees of the Mississippi River, Serial 1572, Washington, 1873.

5. Caleb G. Forshey, "The Levee System of Louisiana," *De Bow's Review* 8 (May 1850): 485.

6. Meier, "Caleb Goldsmith Forshey," 29.

7. Caleb G. Forshey, "(N) Levee Survey and Description of Levee Districts North of Red River, 1851," in *Report of the Senate Standing Committee of 21 March, 1850, on Levees, Drainage, &c* (hereafter *Report on Levees, Drainage*) (New Orleans: Emile Lasere, State Printer, 1853), 24, 34–35.

8. Forshey, "(DDD) Levee Summary," in *Report on Levees, Drainage,* 42.

9. Forshey, "(E) Report of Measurements at Bonnet Carré Crevasse, July 30, 1850," in *Report on Levees, Drainage,* 43–44.

10. Forshey, "(F2) Atchafalaya Basin," in *Report on Levees, Drainage,* 50–51.

11. Forshey, "(G) Report on Bayou Lafourche," in *Report on Levees, Drainage,* 54.

12. Caleb G. Forshey to Andrew A. Humphreys, April 6, 1851, Box 1, Mississippi Delta Survey, Correspondence, 1851–54, Records of the Office of the Chief of Engineers, Record Group 77, National Archives, Washington, D.C. (hereafter MDS).

13. Caleb G. Forshey to Andrew A. Humphreys, Feb. 3, 1859, Box 2, MDS, 1854–59.

14. G. W. R. Bayley to Andrew A. Humphreys, Feb. 27, 1859, Box 2, MDS, 1854–59.

15. Andrew A. Humphreys to G. W. R. Bayley, March 23, 1859, Box 2, MDS, 1854–59.

16. Andrew A. Humphreys to Caleb G. Forshey, March 22, 1859, Box 2, MDS, 1854–59.

17. Caleb G. Forshey to Stephen H. Long, Jan. 13, 1852, Box 1, MDS, 1851–54.

18. Andrew A. Humphreys and Henry L. Abbot, *Report Upon the Physics and Hydraulics of the Mississippi River; Upon the Protection of the Alluvial Region Against Overflow; and Upon the Deepening of the Mouths: Based Upon Surveys and Investigations to Determine the Most Practicable Plan for Securing It from Inundation, and the Best Mode of Deepening the Channels at the Mouths of the River* (1861; Washington, D.C., 1876). Subsequent page references appear in the text.

19. Meier, "Caleb Goldsmith Forshey," 69, 123–24, 134–35.

20. Caleb G. Forshey to Henry L. Abbot, Feb. 13, 1873, Correspondence, vol. 25, Andrew A. Humphreys Papers, Historical Society of Pennsylvania, Philadelphia (hereafter Humphreys Papers).

21. Martin Reuss, "Andrew A. Humphreys and the Development of Hydraulic Engineering: Politics and Technology in the Army Corps of Engineers, 1850–1950," *Technology and Culture* 26 (Jan. 1985): 11.

22. Andrew A. Humphreys to Henry L. Abbot, Feb. 25, 1873, Correspondence, vol. 25, Humphreys Papers.

23. Henry L. Abbot to Caleb Forshey, Feb. 28, 1873, Correspondence, vol. 25, Humphreys Papers.

24. Wooldridge, "Report of the Internal Improvements of Louisiana," Appendix, 9–10; Caleb G. Forshey, "Project of a New Levee System. Drawn Up by Professor Forshey and Submitted by the Senate Committee on Levees—1851," *Report of the Senate*

Committee of 21 March 1850, on Levees, Drainage, &c. (New Orleans: Emile Lasere, 1853), 23–24.

25. Todd Shallat, *Structures in the Stream: Water, Science, and the Rise of the U.S. Army Corps of Engineers* (Austin: University of Texas Press, 1994), 11–42.

26. Burton J. Bledstein, *The Culture of Professionalism: The Middle Class and the Development of Higher Education in America* (New York: Norton, 1976), 88–89.

27. M. C. Clark, "This Committee," *Circular,* Baton Rouge, April 1850, Congress, House Committee on Public Lands, File 31A-G18.8, Record Group 233, National Archives, Washington, D.C., 1.

28. Daniel Hovey Calhoun, *The American Civil Engineer: Origins and Conflict* (Cambridge, Mass.: MIT Press, 1960), 43, 167.

29. Ibid., 190.

30. Raymond H. Merritt, *Engineering in American Society, 1850–1875* (Lexington: University of Kentucky Press, 1969), 29–30, 111.

31. Caleb G. Forshey to Andrew A. Humphreys, Feb. 3, 1859, Box 2, MDS, 1854–59.

32. Legislature of Louisiana, "No. 4. An Act Relative to the Louisiana Levee Company, a Corporation Under the General Laws of this State, Constituting it a Body Politic and Corporate, with Certain Powers, Privileges and Franchises, and Contracting with the Said Corporation for the Construction, Maintenance and Repairs of Certain Levees, and Providing for Compensation Thereof, Feb. 20, 1871," *Acts Passed by the General Assembly of The State of Louisiana* (New Orleans, 1871), 30.

33. Caleb G. Forshey and M. Jeff Thompson, *Report of the Commission of Levee Engineers* (New Orleans, 1873), 5.

34. Ibid., 8, 9.

35. Albert E. Cowdrey, *This Land, This South: An Environmental History* (Lexington: University of Kentucky Press, 1983), 88.

36. Forshey and Thompson, *Report of the Commission of Levee Engineers,* 14.

37. Ibid., 14, 15.

38. Ibid., 40.

39. Ibid., 41, 42.

40. Ibid., 46.

41. Caleb G. Forshey, "Memorial of Citizens of the State of Louisiana," 1. Subsequent page references appear in the text.

42. Caleb G. Forshey, "The Levees of the Mississippi River," *American Society of Civil Engineers. Transactions* 3 (1875): 267–84.

43. Edwin T. Layton, *The Revolt of the Engineers: Social Responsibility and the American Engineering Profession* (Cleveland: Press of Case Western Reserve University, 1971), 30–33, 63; Reuss, "Andrew A. Humphreys and the Development of Hydraulic Engineering," 13; Shallat, *Structures in the Stream,* 189.

44. Albert Stein, "Mississippi Valley. On the Improvement of the River Mississippi," *De Bow's Review* 8 (Feb. 1850): 110.

45. M. Jeff Thompson, *Lecture on the Subject of the Levees Read Before the New Orleans Academy of Sciences* (New Orleans: The Republican Office, 1872), 5.

46. Forshey, "Memorial of Citizens of the State of Louisiana," 1, 15.

Chapter 6. Historical Perspective of Crevasses, Levees, and the Mississippi River

1. P. Viosca, "Louisiana Wetlands and the Value of Their Wildlife and Fishery Resource," *Ecology* 9 (1928): 216–29.

2. For a complete discussion of land loss issues in Louisiana see J. G. Gosselink, C. L. Cordes, and J. W. Parsons, *An Ecological Characterization Study of the Chenier Plain Coastal Ecosystem of Louisiana and Texas,* vol. 1, *Narrative Report,* FWS/OBS-78/9 (Washington, D.C.: U.S. Fish and Wildlife Service, Office of Biological Services, 1979), 302; S. M. Gagliano, K. J. Meyer-Arendt, and K. M. Wicker, "Land Loss in the Mississippi River Deltaic Plain," *Gulf Coast Association of Geological Society Transactions* 17 (1981): 295–306; R. E. Turner and D. R. Cahoon, *Causes of Wetland Loss in the Coastal Central Gulf of Mexico,* vol. 2, *Technical Narrative, Final Report Submitted to Minerals Management Service,* Contract no. 14–12–0001–3252, OCS Study/MMS 87–0120 (Metairie, La.: Minerals Management Service, 1987), 400; and S. Penland, H. F. Roberts, S. J. Williams, A. H. Sallenger Jr., D. R. Cahoon, D. W. Davis, and C. G. Groat, "Coastal Land Loss in Louisiana," *Gulf Coast Association of Geological Society Transactions* 40 (1990): 685–700.

3. Land loss rates have fluctuated over time. Recent decreases in the rate of loss per year are discussed in L. D. Britsch and E. B. Kemp III, *Land Loss Rates: Mississippi River Deltaic Plain,* Technical Report GL-90 (New Orleans: U.S. Army Corps of Engineers, 1990), 2. Although land loss is a recognized coastal issue in south Louisiana, it is also a concern in other parts of the country as well. For a summary of these rates, see J. G. Gosselink and R. H. Baumann, "Wetland Inventories: Wetland Loss Along the United States Coast," *Zeitschrift fur Geomorphologie,* n.f. 34 (1980): 173–87. In Louisiana, the history of land loss is outlined in Gagliano, Meyer-Arendt, and Wicker, "Land Loss in the Mississippi River Deltaic Plain."

4. R. E. Turner, *Relationship Between Canal and Levee Density and Coastal Land Loss in Louisiana,* Biological Report 85(14) (Washington, D.C.: U.S. Fish and Wildlife Service, 1987), 1–6.

5. S. M. Gagliano and J. L. van Beek, *Geologic and Geomorphic Aspects of Deltaic Processes, Mississippi Delta System,* Hydrologic and Geologic Studies of Coastal Louisiana, Report no. 1 (Baton Rouge: Coastal Studies Institute, Louisiana State University, 1970), 140; A. D. Frank, *The Development of the Federal Program of Flood Control on the Mississippi River,* Columbia University Studies in the Social Sciences 323 (New York: Ams Press, 1930), 27, 265; H. D. Vogel, "Annex No. 5, Basic Data Mississippi River," in *Control of Floods in the Alluvial Valley of the Lower Mississippi River,* vol. 1, 1931, 71st Cong., 3rd sess., H. Doc. 798, 61–137.

6. R. H. Kesel, "The Role of the Mississippi River in Wetland Loss in Southeastern Louisiana, U.S.A.," *Environmental Geology and Water Science* 13 (1988): 183–93; J. M. Coleman, "Dynamic Changes and Processes in the Mississippi River Delta," *Geological Society of America Bulletin* 100 (July 1988): 999–1015.

7. F. X. Martin, *The History of Louisiana from the Earliest Period* (New Orleans: J. A. Gresham, 1882; reprint, Gretna, La.: Pelican Publishing, 1995), 469. Also see D. O. El-

liott, *The Improvement of the Lower Mississippi River for Flood Control and Navigation*, vol. 1 (Vicksburg, Miss.: U.S. Army Corps of Engineers, Waterways Experiment Station, 1932), 1–158. This volume provides a detailed discussion of flood and navigation issues along the Mississippi River. To the federal government, navigation was critical and levees were secondary. "Levees were not . . . considered a necessary adjunct to a scheme for navigation improvement" (15).

8. A. A. Humphreys and H. L. Abbot, *Report on the Physics and Hydraulics of the Mississippi River; Upon the Protection of the Alluvial Region Against Overflow; and Upon the Deepening of the Mouths: Based Upon Surveys and Investigations to Determine the Most Practicable Plan for Securing It from Inundation, and the Best Mode of Deepening the Channels at the Mouths of the River* (1861; Washington, D.C., 1876), 556, 168. This treatise was begun with intellectual curiosity, but was intended by Humphreys to be a masterpiece that showcased his reasoning and recommendations and guided U.S. levee policy for nearly 100 years.

9. Morey, *Levees of the Mississippi River*, 42nd Cong., 2d sess., 1872, H. Rept. 44, 1–16.

10. For a complete list of these floods see Humphreys and Abbot, *Report on Physics*, 171–83.

11. Martin, *History of Louisiana*; and Elliott, *Improvement of the Lower Mississippi River*.

12. Peirce F. Lewis, *New Orleans: The Making of an Urban Landscape* (Cambridge: Ballinger, 1976), 64. Situations vary, but as a general rule the drainage system can handle one inch (2.5 cm) of rain the first hour and, as the ground becomes saturated, one-half inch per hour (1.3 cm) thereafter, or three inches (7.6 cm) in five hours. Although it is a well-designed system, in record rainfall events the pumps cannot keep up with the rainfall. For example, if ten inches (25.4 cm) of rain falls in five hours, the system will need about fourteen hours at one-half inch (1.3 cm) per hour to remove the unpumped water, assuming no more rain falls during that fourteen-hour period. Rainfall at a greater rate will probably cause flooding. For a complete account of the accomplishments of A. Baldwin Wood see R. Thompson, "Albert Baldwin Wood—The Man Who Made Water Run Uphill," *New Orleans Magazine* 7, no. 1 (Aug. 1973): 40–43, 74–79.

13. Humphreys and Abbott, *Report on Physics*, 174.

14. C. Ellet Jr., *Report on the Overflows of the Delta of the Mississippi*, 22nd Cong., 1st sess., 1852, S. Exec. Doc. 26, 1–106. Ellet wanted to build outlets (also known as spillways) and reservoirs to decrease the floodwater the river carried. Although probably correct, but counter to the levees-only policy outlined in the Delta Survey and not wanting to be second to anyone, Humphreys demolished Ellet personally for this view. Barry (see n. 29 below) noted Ellet was probably the most renowned engineer in the country. "Charming, athletic, brilliant, handsome, and arrogant, he would risk his own life simply to steal a scene. Ellet had . . . charisma" (36). Also see Martin, *History of Louisiana*; and Elliott, *Improvement of the Lower Mississippi River*.

15. G. Gunter, *The Relationship of the Bonnet Carré Spillway to Oyster Beds in Mississippi Sound and the Louisiana Marsh, with a Report on the 1950 Opening and Study of Beds*

in the Vicinity of the Bohemia Spillway and Baptiste Collette Gap (New Orleans: U.S. Army Corps of Engineers, 1950), 6. For a short summary of flooding through the Bonnet Carré crevasse see *Memorandum Relative to Bonnet Carré* (Vicksburg, Miss.: Mississippi River Commission, 1931), 1–3.

16. Morey, *Levees of the Mississippi River,* 7.

17. A. G. Warfield, *Mississippi River Levees,* 44th Cong., 1st sess., 1876, H. Rept. 494, 8. Further, Gorlinski reports "the inhabitant . . . for over thirty miles down from Lockport suffer annually damages to their crops and lands, such as they cannot often repair, and abandon their homes and property." J. Gorlinski, *Special Report of the Commissioner of the Second Swamp Land District, Submitting Engineer's Report of the Survey of the Valley of Lafourche and Terrebonne to the General Assembly* (Baton Rouge: J. M. Taylor, State Printer, 1859), 5.

18. Humphreys and Abbott, *Report on Physics.* Also, the state engineer reported that in 1849 there were seventeen crevasses south of Lockport that he attributed to the in-filling in Bayou Lafourche associated with the obstructions blocking flow in the Bayou. A. D. Wooldridge, *Appendix Report of the State Engineer* (New Orleans: Emile La Sere, State Printer, 1853), 11.

19. "Crevasse," *Thibodaux Minerva,* Feb. 18, 1854, 2, col. 1.

20. The 1878 General Assembly of the State of Louisiana, at the Second Session of the Fourth Legislature begun and held in the City of New Orleans, Jan. 3, 1876, debated Act No. 9 which urged passage of a bill to remove obstructions placed in Bayou Lafourche by General Jackson in 1814. For more than sixty years the bayou was impassable.

21. See "Crevasse," and "Crevasse at Lockport," *Thibodaux Minerva,* March 25, 1854, 1, col. 1; "All Hopes of Stopping the Crevasse at Lockport Have Been Abandoned," *Thibodaux Minerva,* April 1, 1854, 1, col. 1; and "The Crevasse," *Thibodaux Minerva,* May 6, 1854, 1, col. 5, for a complete discussion of the 1854 flood on Bayou Lafourche.

22. Humphreys and Abbott, *Report on Physics.*

23. A. E. Cowdrey, *Land's End: A History of the New Orleans District, United States Army Corps of Engineers, and Its Lifelong Battle with the Lower Mississippi and Other Rivers Wending Their Way to the Sea* (New Orleans: U.S. Army Corps of Engineers, 1977), 118.

24. According to "The Crevasse at Orleans Plantation," *New Orleans Times,* May 30, 1866, 10, col. 1. Orleans Plantation on the west bank south of Algiers was seven feet deep (2.1 m) through a 75-foot-wide (22.8 m) channel that "inundates the finest portion of the cane crop on the plantation." Further discussion of crevasses during this period can be reviewed in Elliott, *Improvement of the Lower Mississippi River.*

25. *Levees of the Mississippi River,* 51st Cong., 2nd sess., 1891, H. Rept. 3598, 12, 7. See also Annual Report of the Mississippi River Commission for the Fiscal Year Ending June 30, 1890; being Appendix WW of the Annual Report of the Chief of Engineers for 1890 (Washington, D.C.: U.S. Army Corps of Engineers, 1890), 3271.

26. In 1891 Congress began to hear testimony on the condition of the Mississippi River's levees. See *Levees of the Mississippi River. Testimony taken by the Committee on Lev-*

ees and Improvements of the Mississippi River, in Connection with the Bill to Repair and Build the Levees of the Mississippi, to Improve its Navigation, to Afford Ease and Safety to its Commerce, and to Prevent Destructive Flood, 51st Cong., 2d sess., 1891, H. Misc. Doc. 127, 1–72.

27. "Levees," in *Report of the Mississippi River Commission,* Appendix MMM (Washington, D.C.: U.S. GPO, 1912), 3721.

28. *The Problem of the Mississippi River,* 63rd Cong., 1st sess., 1913, S. Doc. 204, 17–28.

29. T. E. Dabney, "Taming the Mississippi," *Jefferson Parish Yearly Review* (1944): 44–48. See J. M. Barry, *Rising Tide: The Great Mississippi Flood of 1927 and How It Changed America* (New York: Simon and Schuster, 1997), 164–65.

30. Ellet, *Report on the Overflows,* 1–106. See also *Mississippi Levees. Memorial of Citizens of the State of Louisiana, in Favor of Nationalizing the Levees of the Mississippi River,* 42nd Cong., 3rd sess., 1873, H. Misc. Doc. 41, 8–9 (much of the material is the same as Morey, *Levees of the Mississippi River);* also A. D. Morehouse, "Reclamation of the Southern Louisiana Wet Prairie Lands," in *Annual Report of the Office of Experiment Stations* (Washington, D.C.: U.S. GPO, 1910), 415–39.

31. Frank, *Development of the Federal Program of Flood Control,* 39.

32. J. Harper, C. Love, D. Bondurant, A. W. Becket, J. O. Farrell, F. Surget, S. Sprague, W. H. Phipps, G. Gillan, D. Walker, H. C. Haley, J. E. Dunn, A. C. Jameson, H. P. Sury, R. Montgomery, F. McKewen, and W. McKewen, *Application for the Construction of Levees on the Mississippi River, and the Reclaiming of Inundated Lands,* 24th Cong., 1st sess., 1836, Am. St. Papers Doc. 1535, 709–11.

33. H. B. Richardson, "Louisiana Levees," in *Riparian Lands of the Mississippi River Past, Present, Prospective,* ed. F. H. Tompkins (New Orleans: Frank M. Tompkins, 1901), 333–44. This is a good summary of the early levee history on the lower Mississippi and the evolution and importance of levee districts.

34. Ellet, *Report on the Overflows.*

35. "Territory Submerged by the Davis Crevasse," *Daily States* [New Orleans], May 22, 1884, 4, col. 3 discusses the possibility that Davis Crevasse may have been a product of an abandoned rice flume.

36. J. Ewens, "Yazoo, Tensas, and Atchafalaya Basins, Flood of 1884," in *Annual Report of the Mississippi River Commission for 1884,* ed. Q. A. Gillmore, C. B. Comstock, C. R. Suter, H. Mitchell, B. M. Harrod, S. W. Ferguson, and R. S. Taylor (Washington, D.C.: Mississippi River Commission, 1885), 102–9. Although a muskrat hole may have been responsible for the break, the old rice flume certainly played an important role. In fact, after the Davis crevasse several bayous were "badly obstructed with logs and overhanging timber" making them impassable for small steamers used by local sugar planters. W. L. Fisk, *Preliminary Examination of Bayou Chevreuil and Bayou Tigre, Louisiana. From Lake Des Allemands to Points near Vacherie, Chigby and Malagay Settlements in St. James Parish for Removal of Bars and Other Obstructions to Navigation,* 52nd Cong., 1st sess., 1891, H. Exec. Doc. 1, Part 2.

37. "Territory Submerged by the Davis Crevasse."

38. See Ewens, "Yazoo, Tensas, and Atchafalaya Basins," 108, where Ewens re-

ports that the old rice flume at the site of the Davis Crevasse had been removed and filled in with the topsoil that could have easily been excavated by burrowing muskrat.

39. G. L. Gillespie, A. Stickney, T. H. Handbury, H. L. Marindin, B. M. Harrod, R. S. Taylor, and H. Flad, "Appendix VV," in *Annual Report of the War Department for the Fiscal Year Ending June 30, 1897, Report of the Chief of Engineers, Part 5*, 55th Cong., 2d sess., 1897, H. Doc. 2, 3505–835. This publication also documents levee costs and those segments constructed in the 1896 to 1897 time period.

40. J. N. O. Wilson, "Report of the Acting Secretary of the Interior, Relative to the Swamp and Overflowed Lands in Louisiana," in *Senate Ex. Doc. 68*, 31st Cong., 1st sess., 1850, 1–8.

41. R. D. Waddill, *History of the Mississippi River Levees, 1717 to 1944: Memorandum.* (Vicksburg, Miss.: Mississippi River Commission, 1945), 1–10.

42. F. Tomlinson, "The Reclamation and Settlement of Land in the United States," *International Review of Agricultural Economics* 4 (1926): 225–72.

43. G. Schneider, *The History and Future of Flood Control* (Chicago: American Society of Civil Engineers, Waterways Division, 1952), 98.

44. In 1828 the federal government reported on the land that was flooded from the Mississippi and suggested that much of this land could be reclaimed. G. Graham, *Act. No. 720, Quantity and Quality of Inundated Lands in Louisiana, the Cost of Reclaiming Them, and Their Value When Reclaimed*, 20th Cong., 2nd sess., 1829, Am. States Papers Doc. 720, 614–18. In 1878 the Louisiana legislature incorporated the Louisiana Land Reclamation Company to assist in the reclamation and protection of land belonging to Louisiana. L. Bush and L. A. Wiltz, "Act No. 52," in *Acts Passed by the General Assembly of the State of Louisiana at the Second Session of the Fifth Legislature, begun and held in the City of New Orleans, January 7, 1878* (New Orleans: Printed at the Office of the Democrat, 1878), 88–91.

45. C. G. Forshey, *Mississippi Levees*, 42nd Cong., 3rd sess., 1873, H. Misc. Doc 41, 1–25. Forshey was a professor of mathematics and engineering and at the time a leading expert on the Mississippi River. See also Morey, *Levees of the Mississippi River.*

46. Waddill, *History of the Mississippi River Levees.* Banks felt during this period that "whoever has control of the water-courses has control of the State [Louisiana]." N. P. Banks, *Suggestions Presented to the Judiciary Committee of the Senate of the U.S. by Major General N. P. Banks, relating to the State of Louisiana*, 38th Cong., 2d sess., 1865, S. Misc. Doc. 9, 6.

47. Frank, *Development of the Federal Program of Flood Control*, 30.

48. Ibid., 39.

49. Waddill, *History of the Mississippi River Levees*, 3, 5.

50. Ellet, *Report on the Overflows.*

51. For an excellent history of the levee builders see Cowdrey, *Land's End.*

52. Waddill, *History of the Mississippi River Levees.*

53. Richardson, "Louisiana Levees," 333–44.

54. F. Simpich, "The Great Mississippi Flood of 1927," *National Geographic* 52 (September 1927): 243–89.

55. Richardson, "Louisiana Levees," 337. Richardson notes further that "there can be no doubt . . . that the work of levee improvement will go on in the future with even more vigor" (344).

56. G. W. Morse, "Report on Bayou Lafourche," *Thibodaux Minerva,* March 17, 1843, 1.

57. Graham, *Act. No. 720,* 614.

58. Morse, "Report."

59. During this period Bayou Lafourche was dammed at its headwaters. This reduced all flow from the Mississippi River. The population began to remove the bayou's levees and cultivate and live on land closer to the bayou. They no longer feared major flooding events. In 1955 water began to flow into Bayou Lafourche from the Mississippi River through a system of pumps.

60. For a complete summary of reclamation activity in Louisiana see C. W. Okey, *The Wetlands of Southern Louisiana and Their Drainage,* U.S. Department of Agriculture, Bulletin no. 71 (Washington, D.C.: U.S. GPO, 1914), 1–82; Robert W. Harrison, *Alluvial Empire: A Study of State and Local Efforts Toward Land Development in the Alluvial Valley of the Lower Mississippi River, Including Flood Control, Land Drainage, Land Clearing, Land Forming,* 2 vols. (Little Rock, Ark.: Pioneer Press, for the Economic Research Service, U.S. Department of Agriculture, 1961), 344; and R. W. Harrison and W. M. Kollmorgan, "Drainage Reclamation in the Coastal Marshlands of the Mississippi River Delta," *Louisiana Historical Quarterly* 30, no. 2 (1974): 654–709.

61. J. O. Wright, "Reclamation of Tide Lands," in United States Department of Agriculture, *Annual Report of the Office of Experiment Stations for the Year Ended June 30, 1906* (Washington, D.C.: U.S. GPO, 1907), 373–96.

62. Barry, *Rising Tide,* 524. This comprehensive analysis of the aftermath of the 1927 flood is an insightful history of the importance of the 1927 flood to American political and economic history.

63. Simpich, "Great Mississippi Flood," 256.

64. A. B. McDaniel, *Flood Waters of the Mississippi River,* 71st Cong., 2d sess., 1930, S. Doc. 127, 1–84. For a map that clearly shows the extent of flooding throughout the Mississippi River's alluvial valley see D. O. Elliott, *The Improvement of the Lower Mississippi River for Flood Control and Navigation,* vol. 3 (Vicksburg, Miss.: U.S. Army Corps of Engineers, Waterways Experiment Station, 1932).

65. Dabney, "Taming the Mississippi," 50. Barry, *Rising Tide,* reports that thirty-nine tons of dynamite was used over a period of ten days to create this crevasse (247).

66. Simpich, "Great Mississippi Flood"; and Elliott, *Improvement of the Lower Mississippi River,* 159–331.

67. A. C. Trowbridge, "Preliminary Geological Report on the Mississippi Delta," Mississippi River Commission, Vicksburg, Miss., 1922, 13.

68. Simpich, "Great Mississippi Flood," 245.

69. Ibid. Barry, *Rising Tide* (photo caption) reports water associated with the Cabin Teele crevasse in Louisiana, near Vicksburg, engulfed the floodplain for 70 miles (112.6 km) to Monroe; as a result 120,000 refugees were added to the Red Cross rolls.

70. Simpich, "Great Mississippi Flood," 264–65.

71. Elliott, *Improvement of the Lower Mississippi River.*

72. J. M. Coleman and H. H. Roberts, "Deltaic Coastal Wetlands," *Geologie en Mijnbouw* 8 (1989): 1–24.

73. R. J. Russell, H. V. Howe, J. H. McGuirt, C. F. Hohm, W. Hadley Jr., F. B. Kniffen, and C. A. Brown, *Lower Mississippi River Delta: Reports on the Geology of Plaquemines and St. Bernard Parishes,* Geological Bulletin no. 8 (New Orleans: Louisiana Geological Survey, 1936), 454.

74. Gagliano, Meyer-Arendt, and Wicker, "Land Loss in the Mississippi River Deltaic Plain."

75. H. Mitchell, *Review of Surveys and Gauging of Cubits Gap, Made in 1868, 1875, and 1876,* 48th Cong., 1st sess., 1883, H. Doc. pt. 2, Report of the Secretary of War, vol. 2, pt. 3, Appendix U of Appendix SS (Mississippi River Commission), 2304.

76. F. A. Welder, *Processes of Deltaic Sedimentation in the Lower Mississippi* (Baton Rouge: Coastal Studies Institute, Louisiana State University, 1959), 90.

77. Coleman, "Dynamic Changes and Processes."

78. See Coleman and Roberts, "Deltaic Coastal Wetlands," and also *Pass a Loutre Crevasse,* 54th Cong. 2d sess., 1897, S. Rept. 1347, 1–6. See also Welder, *Processes of Deltaic Sedimentation.* For a discussion of the changes within the deltaic sequence associated with Garden Island Bay see J. W. Harrier and B. J. Good, "Microfaunal and Sedimentary Changes Associated with a Recent Subdelta Formation of the Mississippi River," *Proceedings Louisiana Academy of Sciences* 53 (1990): 42–52.

79. Russell et al., *Lower Mississippi River Delta.*

80. Coleman, "Dynamic Changes and Processes," and Coleman and Roberts, "Deltaic Coastal Wetlands."

81. D. W. Roberts, J. L. van Beek, S. Fournet, and S. J. Williams, *Abatement of Wetland Loss in Louisiana Through Diversions of Mississippi River Water Using Siphons,* Open-File Report 92–274 (Washington, D.C.: Department of the Interior, U.S. Geological Survey, 1992), 48.

82. J. H. Cowan Jr., R. E. Turner, and D. R. Cahoon, "Marsh Management Plans in Practice: Do They Work in Coastal Louisiana, USA?" *Environmental Management* 12, no. 1 (1989): 37–53. For a detailed and complete history of the design of the Atchafalaya floodway, see M. Reuss, *Designing the Bayous: The Control of Water in the Atchafalaya Basin, 1800–1995* (Alexandria, Va.: Office of History, U.S. Army Corps of Engineers, 1998), 249. Pages 103–10 document crevasses associated with the 1927 floods in the Atchafalaya basin.

83. Roberts et al., *Abatement of Wetland Loss.*

84. Bill Good, Administrator, Coastal Restoration Division, Louisiana Department of Natural Resources, interview with author, March 13, 1999.

85. K. Bahlinger and M. Schrepfer, *1997 Draft Engineering Report, Delta-Wide Crevasses (A Preliminary Report on Existing Crevasses in Delta National Wildlife Refuge and Pass-A-Loutre Wildlife Management Area),* DNR Project No. MR-9 (Baton Rouge: Louisiana Department of Natural Resources, Coastal Restoration Division, 1997), 47.

86. Roberts et al., *Abatement of Wetland Loss.*

87. The development of Louisiana's long-term management plan for the coast is outlined in: Louisiana Coastal Wetlands Conservation and Restoration Task Force, and the Wetlands Conservation and Restoration Authority, *Coast 2050: Towards a Sustainable Coastal Louisiana* (Baton Rouge: Louisiana Department of Natural Resources, 1998), 162. In addition, the Coalition to Restore Coastal Louisiana, *No Time to Lose: Facing the Future of Louisiana and the Crisis of Coastal Land Loss* (Baton Rouge: The Coalition to Restore Coastal Louisiana, 1999), 61, also calls attention to the numerous problems facing Louisiana and the fundamental principles and/or issues involved in restoration policies.

Part 3. Growing Demands of the City

1. For a discussion of attitudes and policies about wetlands see Ann Vileisis, *Discovering the Unknown Landscape: A History of America's Wetlands* (Washington, D.C.: Island Press, 1997).

2. Joel A. Tarr, "The Search for the Ultimate Sink: Urban, Air, Land, and Water Pollution in Historical Perspective," *Records of the Columbia Historical Society* 51 (1984): 1–29.

3. A thorough discussion of lower Mississippi River levee construction appears in Albert E. Cowdrey, *Land's End: A History of the New Orleans District Corps of Engineers* (New Orleans: U.S. Army Corps of Engineers, 1977).

Chapter 7. Perspective, Power, and Priorities

1. Peirce Lewis, *New Orleans: The Making of an Urban Landscape* (Cambridge, Mass.: Ballinger, 1976), 9, 11–16, 54–64; Milburn Calhoun, ed., *Louisiana Almanac, 1997–98* (Gretna, La.: Pelican, 1997), 156. According to the *Louisiana Almanac,* the 1920 U.S. Census recorded 387,219 people in the city of New Orleans.

2. The term *primitive* appeared in a *New Orleans Times-Picayune* article on April 28, 1927. The article is quoted later in this chapter.

3. Calhoun, ed., *Louisiana Almanac, 1997–98,* 154, 156. The U.S. Census for 1920 recorded the population of St. Bernard Parish as 4,968.

4. Harry Hansen, ed., *Louisiana: A Guide to the State,* rev. ed. (New York: Hastings House, 1971), 490–93. For a description of the Isleños' settlements, their history, and their culture, see Gilbert C. Din, *The Canary Islanders of Louisiana* (Baton Rouge: Louisiana State University Press, 1988).

5. Isaac Monroe Cline, *Storms, Floods, and Sunshine* (New Orleans: Pelican, 1945), 197.

6. Pete Daniel, *Deep'n as It Come: The 1927 Mississippi River Flood* (New York: Oxford University Press, 1977), 5–8. The U.S. Mississippi River Commission's "levees only" stance on flood control, in force since 1879, was abandoned in favor of a system of levees and spillways following the devastating flood of 1927. An example of the

clamor for spillways appears in James Parkerson Kemper, *Floods in the Valley of the Mississippi: A National Calamity—What Should Be Done about It* (New Orleans: National Flood Commission, 1928).

7. Daniel, *Deep'n as It Come*, 9–19. A total of 120 crevasses flooded more than 16 million acres in seven states—Missouri, Illinois, Kentucky, Tennessee, Arkansas, Mississippi, and Lousiana. Forty-two major levee breaks occurred in the latter three states, which experienced the worst of the flooding.

Daniel's work includes vivid firsthand accounts from flood victims. A more detailed source on the flood in both the Mississippi Delta and southeast Louisiana is John M. Barry, *Rising Tide: The Great Mississippi Flood of 1927 and How It Changed America* (New York: Simon and Schuster, 1997).

8. Cline, *Storms, Floods, and Sunshine*, 197.

9. Ibid., 198.

10. *New Orleans Times-Picayune*, April 15 through 30, 1927. The first front-page news story on the flood appeared on April 15. Cline may have incorrectly recorded his meeting date with the Association of Commerce, or its censorship may have weakened as a result of Cline's meeting with reporters.

11. *New Orleans Times-Picayune*, April 25, 1927.

12. Hansen, ed., *Louisiana*, 480.

13. Ted O'Neil, *The Muskrat in the Louisiana Coastal Marshes* (New Orleans: Louisiana Department of Wild Life and Fisheries, 1949), 87–90; Robert H. Chabreck, *Coastal Marshes: Ecology and Wildlife Management* (Minneapolis: University of Minnesota Press, 1988), 28–30.

14. R. H. Chabreck and R. E. Condrey, *Common Vascular Plants of the Louisiana Marsh* (Baton Rouge: Louisiana State University Center for Wetland Resources, 1979), 88; O'Neil, *Muskrat in the Louisiana Coastal Marshes*, 64–66, 87. *Ondatra zibethicus rivalicius* is the Louisiana subspecies of the common muskrat.

15. Louisiana Department of Conservation, *Seventh Biennial Report, 1924–1926* (New Orleans: Louisiana Department of Conservation, 1926), 63–66, 72. Louisiana's 1924–25 fur harvest totaled 6,791,265 pelts, of which 6,236,165 were muskrats. In comparison, the entire Dominion of Canada harvested 3,820,326 pelts that season, including 2,515,142 muskrats.

16. O'Neil, *Muskrat in the Louisiana Coastal Marshes*, 68.

17. Ibid.; Louisiana Department of Conservation, *Eighth Biennial Report, 1926–1928* (New Orleans: Louisiana Department of Conservation, 1928), 299–304.

18. *New Orleans Times-Picayune*, April 22 through 25, 1927. The crevasses that gave rise to these alarming headlines occurred on April 21 at Mount (Stops) Landing, Mississippi, and Pendleton, Arkansas. The breaks quickly flooded Greenville and Arkansas City, along with surrounding rural areas.

19. *New Orleans Times-Picayune*, April 25, 1927; Cline, *Storms, Floods, and Sunshine*, 199.

20. Cline, *Storms, Floods, and Sunshine*, 199–200; Lyle Saxon, *Father Mississippi* (New York: Century, 1927), 322–28. Saxon's book was published in Oct. 1927 and is filled with accounts of residents' anxiety as the flood crest approached New Orleans.

21. Saxon, *Father Mississippi*, 326–29; Cline, *Storms, Floods, and Sunshine*, 200.

22. Barry, *Rising Tide*, 247, 250.

23. *New Orleans Times-Picayune*, April 27 (headline) and April 28 and 29, 1927; Saxon, *Father Mississippi*, 330–33. The refugee camp was located in the eastern section of the city at Poland Street and the Mississippi River.

24. *New Orleans Times-Picayune*, April 28, 1927.

25. The "trapper's war" was an armed uprising in the marshes of St. Bernard Parish in Nov.–Dec. 1926. The uprising occurred in response to conflicting claims of landowners and trappers, such as the landowners' new policy that fur trappers lease their trapping grounds or pay a percentage of their earnings to the landowner; see Louisiana Department of Conservation, *Seventh Biennial Report*, 106–11; *Eighth Biennial Report*, 298; and *Thirteenth Biennial Report, 1936–37* (New Orleans: Louisiana Department of Conservation, 1938), 57. *New Orleans Times-Picayune*, April 28, 1927.

26. Joseph Valsin Guillotte III, *Masters of the Marsh: An Introduction to the Ethnography of the Isleños of Lower St. Bernard Parish, Louisiana* (N.p.: Jean Lafitte National Historical Park, c. 1982), 115.

27. For information on the fur industry in south Louisiana, see Gay M. Gomez, *A Wetland Biography: Seasons on Louisiana's Chenier Plain* (Austin: University of Texas Press, 1998), 140–61.

28. Saxon, *Father Mississippi*, 337. Saxon is quoting an interview by newspaper reporter Meigs O. Frost of the *New Orleans States*.

29. Ibid., 339. The blasting is described in the *New Orleans Times-Picayune*, April 30, 1927. The "public execution" of Meraux's parish may have been unnecessary, for the day after the first dynamite blast a crevasse in the Glasscock levee upriver in Concordia Parish released Mississippi River floodwaters into the Atchafalaya Basin.

30. *New Orleans Times-Picayune*, May 1 through 10, 1927. Each day the newspaper reported donations of food, clothing, accommodations, and prayers by the city's grateful residents; it also reported the organization of a reparations committee to reimburse flood victims for their losses. These initial concerns for the refugees dimmed during the following three months, however, as costs of food and housing mounted to over $20,000 per week. When the Red Cross refused to aid the city in caring for the refugees, the reparations committee voted to deduct the costs of food and shelter from the victims' claim settlements (Barry, *Rising Tide*, 247).

31. *New Orleans Times-Picayune*, May 2, 1927.

32. Ibid., May 1 and 2, 1927; Louisiana Department of Conservation, *Eighth Biennial Report*, 240–42. Such a slaughter occurred in 1922 when the levee broke at Poydras; faced with the loss of an important part of their income that spring, trappers killed and sold nearly 100,000 thin-furred "summer muskrats."

33. O'Neil, *Muskrat in the Louisiana Coastal Marshes*, 87.

34. Louisiana Department of Conservation, *Eighth Biennial Report*, 233–34. Subsequent page references appear in the text.

35. Ibid., 323; O'Neil, *Muskrat in the Louisiana Coastal Marshes*, 68. According to the *Eighth Biennial Report*, the parish's oyster industry would recover in just over two years as well; the Department of Conservation predicted a "banner crop" in the 1929–30

season, despite the loss of one-third to one-half of the parish's oysters in the flood of 1927.

36. Louisiana Department of Conservation, *Eighth Biennial Report*, 321–22; Barry, *Rising Tide*, 356–59. Barry's *Rising Tide* includes a detailed account of wrangling by the banker-controlled Reparations Committee to reduce the city's claim payments. The $2.9-million settlement went primarily to six large claimants, leaving approximately $800,000 to divide among 2,809 claimants. Each received an average payment of $284, while an additional 1,024 claimants received nothing at all.

37. Hansen, ed., *Louisiana*, 488.

38. Barry, *Rising Tide*, 408.

39. I refer primarily to damage inflicted by the Mississippi River Gulf Outlet, a seventy-six-mile navigation channel that slashes through the marshes of St. Bernard Parish, providing a shortcut route for oceangoing vessels to reach the Port of New Orleans (Hansen, ed., *Louisiana*, 311). Completed in 1963, the channel has greatly facilitated the movement of saltwater into the area's wetland ecosystems, a factor believed by both St. Bernard's residents and wetland ecologies to be contributing to coastal land loss and declining productivity in the region (Guillotte, *Masters of the Marsh*, 115–16; Coalition to Restore Coastal Louisiana, *No Time to Lose: Facing the Future of Louisiana and the Crisis of Coastal Land Loss* [Baton Rouge: Coalition to Restore Coastal Louisiana, 1999], 18–23).

Chapter 8. In the Wake of Hurricane Betsy

1. Mark Monmonier, *The Cartographies of Danger: Mapping Hazards of America* (Chicago: University of Chicago Press, 1997), 69–70; for hurricane horsepower equivalents see Paul Simons, *Weird Weather* (Boston: Little, Brown, 1996), 99; and Robert Henson, "The Intensity Problem," *Weatherwise* (Sept./Oct. 1998): 20–26.

2. Michael Pafit, "Living with Natural Hazards," *National Geographic* (July 1998): 12.

3. U.S. House, Committee on Public Works, "Hearings before the Special Subcommittee to Investigate Areas of Destruction of Hurricane Betsy," 89th Cong., 1st sess. (1965), 10; see also Albert E. Cowdrey, *Land's End: A History of the New Orleans District, U.S. Army Corps of Engineers* (New Orleans: U.S. Army Corps of Engineers, 1977), 49, 78–79.

4. U.S. Army Engineer District, New Orleans, *History of Hurricane Occurrences Along Coastal Louisiana* (New Orleans: U.S. Army Corps of Engineers, 1972), 3, 12; for the technology of warning and evacuation see, for example, "At War with the Weather," *Science News*, July 9, 1966, 26–27; see also, Shari Rudavsky, "Technology Brings Forecaster's Eyes Closer to Eye of the Storm," *Washington Post*, Aug. 31, 1992; for human-induced disasters see Edward Bryan, *Natural Hazards* (Cambridge: Cambridge University Press, 1991), passim.

5. Edward Tenner, *Why Things Bite Back* (New York: Vintage Books, 1996), 91; see also Ian Burton, Robert W. Kates, and Gilbert F. White, *The Environment as Hazard*, 2nd ed. (New York: Guilford Press, 1993), 17–18, 34–35; and Joseph T. Kelley, Alice R.

Kelley, Orrin H. Pikey Sr., and Albert A. Clark, *Living with the Louisiana Shore* (Durham, N.C.: Duke University Press, 1984), 4–9.

6. "Bouncing Betsy," *Newsweek,* Sept. 20, 1965, 25; U.S. House, Committee on Public Works, "Destruction of Hurricane Betsy," 29, 467.

7. U.S. House, Committee on Public Works, "Destruction of Hurricane Betsy," 25.

8. Ibid., 81, 111.

9. "Betsy's Toll," *Newsweek,* Sept. 27, 1965.

10. Joe Englert, ed., *The Great Disaster: Hurricane "Betsy," Sept. 9, 1965* (Chalmette, La.: Kaiser Aluminum, 1965), 1–2; see also "Hurricane Pounds at New Orleans; 185,000 Flee in Southern Louisiana," *New York Times,* Sept. 10, 1965; and Roy Reed, "New Orleans Loss in Storm Heavy; 23 Dead in 3 States," *New York Times,* Sept. 11, 1965.

11. U.S. House, Committee on Public Works, "Destruction of Hurricane Betsy," 188–90.

12. Ibid., 121; see also Englert, ed., *Great Disaster,* 1

13. Englert, ed., *Great Disaster,* 1; see also U.S. Army Corps of Engineers, *Report on Hurricane Betsy* (New Orleans: Corps of Engineers, 1965), 1–13, 32.

14. "The 'New' New Orleans—Comeback of a Southern City," *U.S. News and World Report,* July 25, 1966, 80; see also George Laycock, *The Diligent Destroyers* (New York: Doubleday, 1970), 167, 172–73, 181.

15. Frank "Big Kenny" Campo and Frank "Blackie" Campo, interview with author, Shell Beach, La., July 10, 1996.

16. Lyle Saxon, *Father Mississippi* (New York: Century, 1927), 324. The story of Old Shell Beach is based on a series of interviews with Frank "Kenny" Campo and Frank "Blackie" Campo at new Shell Beach, Louisiana, 1996–98; see also Joseph Guillotte, interview with author, University of New Orleans, Aug. 27, 1997; Harnett T. Kane, *The Bayous of Louisiana* (New York: William Morrow, 1943), 112; Gilbert C. Din, *The Canary Islanders of Louisiana* (Baton Rouge: Louisiana State University Press, 1988), 126, 129; and Glen Jeansonne, *Leander Perez: Boss of the Delta* (Baton Rouge: Louisiana State University Press, 1977), 32–34.

17. *1927 Edition of the Sears, Roebuck Catalogue* (New York: Bounty Books, 1970), 18; see also Abe Cochran, "New Directions for the Fur Industry," *Louisiana Conservationist* (March/April 1991): 26–29.

18. William Chase to Charles Gratiot, Feb. 9, 1837, *House Exec. Doc. 173,* 24th Cong., 2nd sess., p. 5.

19. "Statement of Hon. Russell B. Long," *Mississippi River-Gulf Outlet,* House Report No. 6309, 8th Cong., 2nd sess. (Jan. 19 and 20, 1956), 5.

20. Bob Anderson, "Navigation Channel Taking Toll on Marsh," *Baton Rouge Morning Advocate,* May 13, 1991; see also Earth Search, Inc., *Cultural Resources Survey of the Mississippi River Gulf Outlet Dredged Material Disposal Areas, St. Bernard Parish, Louisiana* (New Orleans: U.S. Army Corps of Engineers, 1993), 63.

21. Harley Winer, conversation with author, Harahan, La., Oct. 16, 1998.

22. Mark Schleifsein and Keith Darce, "Goodbye Mister Go: Economics, Envi-

ronment May Doom Shipping Channel," *New Orleans Times-Picayune,* Oct. 18, 1998; see also Lloyd Lee Gregory, "Sinking the Swamp or MRGO, a Canal Only an Engineer Could Love: A Critical Assessment of the Mississippi River-Gulf Outlet," New Orleans, Tulane Environmental Law Clinic, n.d., 3.

23. Robert Guizerix, Chief, Structures Branch, U.S. Army Corps of Engineers New Orleans District, interview with author, June 19, 1996, and Nov. 18, 1998.

24. Cecil Soileau, interview with author, New Orleans, Dec. 29, 1998.

25. William Ophuls, *Ecology and the Politics of Scarcity* (San Francisco: W. H. Freeman, 1977), 176–77; see also Richard N. Andrews, *Environmental Policy and Administrative Change: Implementation of the National Environmental Policy Act* (Lexington, Mass.: D. C. Heath, 1976).

26. Charles "Pete" Savoye, interview with author, Chalmette, La., June 19, 1996; for the port generally see U.S. Board of Engineers for Rivers and Harbors, *The Port of New Orleans, Louisiana,* 2 parts (Washington, D.C.: USGPO, 1971), 2:2, 10–12; Cowdrey, *Land's End,* 70; and, Helen Delich Bentley, ed., *Ports of the Americas: History and Development* (N.p.: American Association of Port Authorities, 1961), 120–24.

27. Col. Herbert R. Haar Jr., "Centroport—for the Year 2000," *Military Engineer* 65 (Jan.–Feb. 1973): 1; for the Corps' response to NEPA see Ophuls, *Ecology and the Politics of Scarcity,* 176–77; Walter A. Rosenbaum, *The Politics of Environmental Concern,* 2nd ed. (New York: Praeger, 1977), 188; Council on Environmental Quality, *Environmental Quality: Sixth Annual Report* (Washington, D.C.: Council on Environmental Quality, 1975), 636–37; and Martin Reuss, *Shaping Environmental Awareness: The United States Army Corps of Engineers Environmental Advisory Board, 1970–1980* (Washington, D.C.: U.S. Army Corps of Engineers Historical Division, 1983), 7–16.

28. Charles "Pete" Savoye even built a model of Mister Go with a two-inch plastic dinosaur. "That's the corps," he explained in a June 16, 1997, backyard interview in Chalmette.

29. "Hebert Pushes for Lock, Canal," *New Orleans Times-Picayune,* Jan. 26, 1973; see also Daniel A. Mazmanian and Jeanne Nienaber, *Can Organizations Change?: Environmental Protection Citizen Participation, and the Corps of Engineers* (Washington, D.C.: The Brookings Institution, 1979), 92; and Sue Hawes, Corps of Engineers New Orleans District, interview with author, Oct. 24, 1996 (hereafter Hawes interview).

30. Hawes interview, Oct. 24, 1996.

31. U.S. Army Corps of Engineers, *Digest of Water Resources Policies and Authorities,* EP 1165–2–1, Feb. 15, 1996, appendix B; see also Daniel D. Cook, "Will the Army Corps Engineer a New Image?" *Industry Week* 199 (Oct. 2, 1978): 34–39; and Martin Heuvelman, *The River Killers* (Harrisburg, Pa.: Stackpole Books, 1974), 127, 213–17.

32. David Zwick and Marcy Benstock, *Water Wasteland: Ralph Nader's Study Group Report on Water Pollution* (New York: Grossman, 1971), 19–20, passim; see also Robert W. Adler, Jessica C. Landman, and Diane M. Cameron, *The Clean Water Act 20 Years Later* (Washington, D.C.: Island Press, 1993), 5–6.

33. From the legislative history of the act, as quoted in Edward Thompson Jr., "Section 404 of the Federal Water Pollution Control Act Amendments of 1977: Hy-

drologic Modification, Wetland Protection and the Physical Integrity of the Nation's Waters," *Harvard Environmental Law Review* 2 (1977): 272.

34. L. Kent Brown, Corps of Engineers New Orleans District, interview with author, Boise, Idaho, Nov. 22, 1996, and Jan. 2, 1997.

35. U.S. Fish and Wildlife Service, *Mississippi Gulf Outlet, Ship Lock and Connecting Channels Study (Planning-Aid Report)* (Washington, D.C.: U.S. Fish and Wildlife Service, March 1982), appendix F.

36. Mazmanian and Nienaber, *Can Organizations Change?* 97; see also "St. Bernard Ship-Barge Lock Gains Endorsement," *New Orleans Times-Picayune,* Oct. 28, 1972; see also Cornelia Carrier, "Marshlands Key 1972 Environmental Concern," *New Orleans Times-Picayune,* Jan. 28, 1973; and Edgar Poe, "Corps Urged to Hurry Up with Ship Lock," *New Orleans Times-Picayune,* Feb. 3, 1978.

37. The exchange between Senator Long and LBJ is reported in U.S. House, Committee on Public Works, "Destruction of Hurricane Betsy," 130.

38. "President Tours Hurricane Area," *New York Times,* Sept. 11, 1965.

39. Cowdrey, *Land's End,* 80; see also U.S. Army Corps of Engineers, *Hurricane Betsy, September 8–11, 1965* (New Orleans: Corps of Engineers, 1965), passim; and Public Law 89–298, Oct. 27, 1965.

40. Fred Chatry of the Corps' New Orleans District provided a helpful chronology of events; see also Hawes interview, Nov. 1, 1996; Barry Kohl, New Orleans Audubon Society, interview with author, June 13, 1996; and Cornelia Carrier, "Expert: Lake's Productivity in Danger," *New Orleans Times-Picayune,* Nov. 19, 1974.

41. John Fahey, "Pontchartrain: Cesspool or Playground? Environmentalists, Dredgers Differ," *New Orleans Times-Picayune,* Jan. 23, 1975; Luke Fontana, Save Our Wetlands, Inc., interview with author, New Orleans, Nov. 4–6, 1996; "Dredging Foes Win One Round," *New Orleans Times-Picayune,* Nov. 1, 1974; see also Cornelia Carrier, "Wetlands Dispute New Trial Denied," *New Orleans Times-Picayune,* Feb. 21, 1975; Cornelia Carrier, "Lake Plan Hit as Private Windfall," *New Orleans Times-Picayune,* Feb. 23, 1975; Gordon Gsell, "SOWL Files Suit against Lake Plan," *New Orleans Times-Picayune,* Dec. 6, 1975; John LaPlace, "Could a State Panel Save Pontchartrain?" *New Orleans Times-Picayune,* March 25, 1977; Cornelia Carrier, "La. Delegation Is Environment's Foe," *New Orleans Times-Picayune,* June 15, 1977; and "Environmental Group Asks to Join Hurricane Plan Suit," *New Orleans Times-Picayune,* Dec. 7, 1977.

42. Glen Montz, Corps of Engineers New Orleans District, interview with author, New Orleans, Nov. 18, 1996; Milton Cambree, St. Charles Parish, interview with author, June 15, 1997; and U.S. Army Corps of Engineers, *Lake Pontchartrain, Louisiana, and Vicinity Hurricane Protection Project* (New Orleans: U.S. Army Corps of Engineers, 1984), 2; "Final Environmental Impact Statement Available," *New Orleans Times-Picayune,* Jan. 19, 1975; "Hurricane Plan Support Falters," *New Orleans Times-Picayune,* Jan. 6, 1978; J. Douglas Murphy, "Flood Barriers, Floodgates at Rigolet, Chef Pass 'Best Plan'—Meteorologist," *New Orleans Times-Picayune,* Jan. 30, 1975; and Edgar Poe, "Lake Pontchartrain Flood Aid $21 Million," *New Orleans Times-Picayune,* Dec. 12, 1975.

43. "Reconsidering the Barriers," *New Orleans Times-Picayune,* Jan. 4, 1978; see also

"Lake Study Falls Short," *New Orleans Times-Picayune*, Jan. 30, 1978; U.S. General Accounting Office, *Report to the Secretary of the Army: Improved Planning Needed by the Corps of Engineers to Resolve Environmental, Technical, and Financial Issues on the Lake Pontchartrain Hurricane Protection Project*, GAO/MASAD-82–39 (Washington, D.C.: General Accounting Office, 1982), 5.

44. Hawes interview, Nov. 1, 1996; Joseph A. Towers, Corps of Engineers New Orleans District, New Orleans, Nov. 6, 1996; Robert J. Guizerix, Corps of Engineers New Orleans District, interview with author, New Orleans, June 19, 1996.

45. Bonnie S. Lawrence, ed., *Restless Earth* (Washington, D.C.: National Geographic Society, 1997), 140–41.

46. Anders Wijkman and Lloyd Timberlake, *Natural Disasters: Acts of God or Acts of Man* (London: Earthscan, 1984), 11; Burton, Kates, and White, *Environment as Hazard*, 81–84; see also Peter Applebome, "Storm Cycles and Coastal Growth Could Make Disaster a Way of Life," *New York Times*, Sept. 2, 1992; Donna Abu-Nasr, "Disaster after Disaster, 1988's Been a Costly Year," *Idaho Statesman*, Nov. 28, 1998; and "Is the United States Headed for Hurricane Disaster?" *Bulletin of the American Meteorological Society* 67 (May 1986): 537–38.

47. Cory Dean, "Louisiana Wetlands, a Critical Habitat Founders," *New York Times*, Nov. 20, 1990; see also Oliver A. Houck, "Land Loss in Louisiana: Causes, Consequences, and Remedies," *Tulane Law Review* 58, no. 1 (1983): 3–168.

48. Rutherford H. Platt, Timothy Beatley, and H. Crane Miller, "The Folly at Folly Beach and Other Failings of U.S. Coastal Erosion Policy," *Environment* 33 (Nov. 1991): 6–9, 25–32; for the MRGO critique see, for example, Amy Mathews-Amos, ed., *Costly Corps: How the U.S. Army Corps of Engineers Spends Your Tax Dollars to Destroy America's Natural Resources* (New Orleans: Gulf Restoration Network and Sierra Club Legal Defense Fund, 1996), 20–22; the counterargument (that the MRGO provides 12,000 jobs and containership service while its dredge material builds wetlands) is increasingly hard to make in the wake of 1998 Hurricane George that shifted a shoal in front of the mouth of the channel; see Julie Aitken, "MRGO Sediment Builds Wetlands," *Riverside* (March 1995): 4; and Timothy P. Ryan, "The Economic Impact of Closing the Mississippi River Gulf Outlet," draft, prepared for the Port of New Orleans, 1993.

49. Robert D. Brown, "Restoring Louisiana's Wetlands," *Military Engineer*, no. 550 (July 1992): 4.

Part 4. Response to Environmental Change

1. Ian Douglas, *The Urban Environment* (London: Edward Arnold, 1983).

Chapter 9. Too Much of a Good Thing

Note: I would like to thank my colleague John Tiefenbacher for reviewing a draft of this paper, Stephanie Garcia and Joy Adams for their excellent research assistance, and Paul Schultze for preparing the maps. Comments from anonymous reviewers also improved this chapter.

1. While little history has been written on pollution control on the lower river, there has been considerable discussion of the upper river and its tributaries. See Philip Scarpino, *The Great River: An Environmental History of the Upper Mississippi, 1890–1950* (Columbia: University of Missouri Press, 1985); Joel A. Tarr, "Searching for a 'Sink' for an Industrial Waste: Iron-Making Fuels and the Environment," *Environmental History Review* 18 (Spring 1984): 9–34; and Craig E. Colten, "Illinois River Pollution Control, 1900–1970," in *The American Environment: Interpretations of Past Geographies*, ed. Lary Dilsaver and C. E. Colten (Lanham, Md.: Rowman and Littlefield, 1992), 193–216. These works acknowledge dilution as a primary pollution-control method in the early twentieth century, but illustrate that upstream officials took action to restrict industrial discharges long before Louisiana sought to moderate pollution on the lower Mississippi.

2. James C. Cobb, *Industrialization and Southern Society, 1877–1984* (Lexington: University of Kentucky Press, 1984), 40. See also Philip Scranton, ed., *Second Wave: Southern Industrialization, 1940–1970* (Athens: University of Georgia Press, forthcoming); U.S. Public Health Service (hereafter USPHS), *Summary Report on Water Pollution: Southwest-Lower Mississippi Drainage Basins* (Washington, D.C.: USPHS, Division of Water Pollution, 1951); U.S. Department of Commerce, Bureau of Census, *Census of Manufacturers, 1947*, vol. 3, *Statistics by States* (Washington, D.C.: Bureau of Census, 1950); *Census of Manufacturers, 1954*, vol. 3, *Area Statistics* (1955); *Census of Manufacturers, 1958*, vol. 3, *Area Statistics* (1961); *1963 Census of Manufacturers*, vol. 3, *Area Statistics* (1965); *1967 Census of Manufacturers*, vol. 3, *Area Statistics* (1971).

3. U.S. Environmental Protection Agency (hereafter USEPA), *Industrial Pollution of the Lower Mississippi River* (Dallas: USEPA, 1972).

4. Cobb, *Industrialization and Southern Society*, and *The Selling of the South: The Southern Crusade for Industrial Development, 1936–1980* (Baton Rouge: Louisiana State University Press, 1982); William B. Meyer, "Decentralized Regulation and the 'Race to the Top': The Case of Municipal Salon Licensing in Turn-of-the-Century Massachusetts," *Professional Geographer* 44, no. 3 (Aug. 1992): 286–95.

5. Tarr, "Searching for a 'Sink.'" Concern with oil pollution led to federal studies of oilfield and refinery wastes in the 1920s but concentrated on coastal waters. See U.S. Department of the Interior, Bureau of Mines, *Pollution by Oil of the Coast Waters of the United States: Preliminary Report* (Washington, D.C.: U.S. Department of the Interior, 1923). This report found oilfield pollution affecting the Gulf Coast but did not investigate navigable rivers such as the Mississippi. Illinois officials began restricting oil releases to the upper Mississippi in the 1930s. "Abatement of Industrial Pollution in Illinois," *Public Works* 67 (1936): 18.

6. E. J. Cleary, "Determining Risks of Toxic Substances in Water," *Sewage and Industrial Wastes* 25 (1954): 203–10; and Jules S. Cass, "The Potential Toxicity of Chemicals in Water for Man and Domestic Animals," *Proceedings of the Tenth Industrial Waste Conference* (Lafayette, Ind.: Purdue University, 1955), 466–72.

7. *McFarlain v. Jennings-Heywood Oil Syndicate et al.* (1907) 118 La. 537; *Long v. Louisiana Creosoting* (1916) 69 S. 281; *Williams et al. v. Pelican Natural Gas Co. et al.* (1937) 1 75 S. 28; *Connell et al. v. International Paper* (1951) 99 F. Supp. 699.

8. Ibid. Christine Rosen has argued that pollution abatement technology allowed judges in other states to order injunctions against polluters. Their actions, she suggests, were intended to force manufacturers to reduce pollution, not close their operations. Christine Rosen, "Differing Perceptions of the Value of Pollution Abatement Across Time and Place," *Law and History Review* 11 (1993): 303–81.

9. "Act 183: An Act to Protect the Rice Planters . . .," *Acts Passed by the General Assembly of the State of Louisiana, 1910* (Baton Rouge: New Advocate, 1910), 272–73.

10. "Act 133: An Act to Protect All the Natural Waterways and Canals . . .," *Acts Passed by the Legislature of the State of Louisiana, 1924* (Baton Rouge: Ramires-Jones Printing, 1924), 199–200.

11. *State of Louisiana v. Hincy* (1912), 58 S. 41 1.

12. "Act 367: An Act to Create a Stream Control Commission," *Acts Passed by the Legislature of the State of Louisiana, 1940* (Baton Rouge, 1940), 1369–74; and Louisiana Department of Conservation, *Fifteenth Biennial Report, 1940–41* (Baton Rouge: Louisiana Department of Conservation, 1940–41), 139–41.

13. Pratt has argued that the petrochemical industry sought hegemony over the pollution-control process by seeking to establish acceptable treatment methods. Joseph Pratt, "Letting the Grandchildren Do It: Environmental Planning During the Ascent of Oil as a Major Oil Source," *Public Historian* 2, no. 4 (Summer 1980): 28–61. Colten has found that the petrochemical industry worked closely with state officials to determine what constituted pollution and even took on environmental monitoring responsibilities in Texas. Craig E. Colten, "Texas v. the Petrochemical Industry: Contesting Pollution in an Era of Industrial Growth," in *The Second Wave*, ed. Scranton.

14. Louisiana Department of Wild Life and Fisheries, *Second Biennial Report* (Baton Rouge: Louisiana Department of Wild Life and Fisheries, 1946–47), 328–29.

15. Louisiana Department of Wild Life and Fisheries, *Third Biennial Report, 1948–49* (Baton Rouge: Louisiana Department of Wild Life and Fisheries, 1948–49), 352–53.

16. Louisiana Department of Wild Life and Fisheries, *First Biennial Report, 1944–45* (New Orleans: Louisiana Department of Wild Life and Fisheries, 1946), 218.

17. Ibid.

18. Louisiana Department of Conservation, *Sixteenth Biennial Report, 1942–43* (New Orleans: Louisiana Department of Conservation, 1944), 172–73.

19. Louisiana Department of Conservation, *Second Biennial Report, 1946–47* (New Orleans: Louisiana Department of Conservation, 1948), 348–49.

20. Louisiana Department of Wild Life and Fisheries, *Third Biennial Report, 1948–49*, 368–69.

21. "Act 367: An Act to Create a Stream Control Commission"; *Texas Company v. Montgomery* (1948) 73 F. Supp. 527.

22. Louisiana Department of Conservation, *Fifteenth Biennial Report, 1940–41*, 136;

and *Sixteenth Biennial Report, 1942–43*, 166–87; Louisiana Board of Health, *Biennial Report* (New Orleans: Louisiana Board of Health, 1942–43), 171–72.

23. USPHS, *Summary Report on Water Pollution: Southwest-Lower Mississippi Drainage Basins* (Washington, D.C.: USPHS, Division of Water Pollution, 1951), 132–33.

24. Frank W. MacDonald, "An Investigation of the Pollution of the Mississippi River in the Vicinity of New Orleans," dissertation, Tulane University, 1953, 65; USPHS, *Summary Report on Water Pollution*, 132.

25. U.S. Department of Health, Education, and Welfare (hereafter DHEW), *1957 Inventory: Municipal and Industrial Waste Facilities* (Washington, D.C.: USHEW, 1958). Louisiana compiled its data in 1953.

26. M. L. Eddards, L. R. Kister, and Glenn Scarcia, *Water Resources of the New Orleans Area, Louisiana* (Washington, D.C.: U.S. Geological Survey, Geological Survey Circular 374, 1956).

27. Louisiana Stream Control Commission, *Proceedings of Meeting* (Baton Rouge, May 15, 1958–Sept. 22, 1966). Housed at the Louisiana Department of Environmental Quality, Baton Rouge.

28. Louisiana Stream Control Commission, *Proceedings of Meeting* (Baton Rouge: June 30, 1961) 15–25. The commission considered an application by California Chemical to release effluent containing high concentrations of phenol and salts to the river and effectively placed the application on hold until more information was assembled. No subsequent permit application appears in the somewhat incomplete record for this plant.

29. Louisiana Stream Control Commission, *Proceedings of Meeting* (Baton Rouge: May 15, 1958–Sept. 22, 1966). Also the SCC reported an approximate 25 percent increase in chlorides in 1960. Louisiana Stream Control Commission, *Proceedings of Meeting* (Baton Rouge: Feb., 16, 1960), 10.

30. Louisiana Wild Life and Fisheries Commission, *Eighth Biennial Report, 1958–59* (New Orleans, 1960), 170–71.

31. Ibid., 184.

32. Louisiana Wild Life and Fisheries Commission, *Ninth Biennial Report, 1960–61* (New Orleans, 1962), 203–4.

33. Louisiana Stream Control Commission, *Proceedings of Meeting* (Baton Rouge: April 5, 1960), 2–3.

34. Louisiana Wild Life and Fisheries Commission, *Ninth Biennial Report,' 1960–61* (New Orleans, 1962), 197–98.

35. Louisiana Wild Life and Fisheries Commission, *Eighth Biennial Report, 1958–1959* (New Orleans, 1960), 187; U.S. Department of the Interior, Federal Water Pollution Control Administration, *Endrin Pollution in the Lower Mississippi River Basin* (Dallas: U.S. Department of the Interior, Federal Water Pollution Control Administration, 1969), 61.

36. Kenneth E. Biglane, "Statement," *Proceedings: Conference in the Matter of Pollution of the Interstate Waters of the Mississippi River* (Washington, D.C.: DHEW, May 5–6, 1964), 1:20–21.

37. See "Poisons Kill Fish in the Mississippi," *New York Times,* March 22, 1964, 79; other articles in 1964 appeared on March 24, p. 32; April 7, p. 55; April 8, p. 23; April 10, p. 27; April 23, p. 41; April 24, p. 35; April 26, p. 60; and May 6, p. 52.

38. USPHS, "Report on Investigation of Fish Kills in Lower Mississippi River and Atchafalaya River, and Gulf of Mexico" (Washington, D.C.: DHEW, 1964), reprinted as exhibit 160 in U.S. Senate, Committee on Government Operations, *Interagency Coordination in Environmental Hazards (Pesticides),* 88th Cong., 2nd sess., April 7, 8, and 15, 1964, 1697.

39. Ibid., 1698–1701.

40. Federal Water Pollution Control Administration, *Endrin Pollution in the Lower Mississippi River Basin,* 71.

41. Alfred D. Grzenda, "Statement," *Conference in the Matter of Pollution of the Interstate Waters of the Mississippi River: Proceedings* (Washington, D.C.: DHEW, May 5–6, 1964), 2:182–258.

42. Bernard Lorant (Vice President, Velsicol Chemical Corporation), "Statement," in *Conference in the Matter of Pollution of the Interstate Waters of the Lower Mississippi River: Proceedings* (Washington, D.C.: DHEW, May 5–6, 1964), 3:487–91.

43. Federal Water Pollution Control Administration, *Endrin Pollution in the Lower Mississippi River Basin.*

44. A chemical trade magazine report concluded there was no threat to humans. See Warren Kornbert, "So What Did Kill the Fish," in U.S. Senate, Committee on Government Operations, *Interagency Coordination in Environmental Hazards (Pesticides),* 1879–89. Originally printed in *Chemical Week* 95, no. 4 (July 27, 1964): 5, 19–26.

45. Louisiana Board of Health, *Biennial Report, 1964–65* (New Orleans, 1965), 105.

46. Louisiana Board of Health, *Biennial Report, 1966–67* (New Orleans, 1967), 92.

47. Lucia Dunham, Roger O'Gara, and Floyd Taylor, "Studies on Pollutants from Processed Water: Collection from Three Stations and Biologic Testing for Toxicity and Carcinogenesis," *American Journal of Public Health* 57, no. 12 (1967): 2178–85.

48. James Friloux, "Petrochemical Wastes as a Water Pollution Problem in the Lower Mississippi River," unpublished paper submitted to the Senate Subcommittee on Air and Water Pollution, April 5, 1971 (from the personal collection of the author).

49. USEPA, *Industrial Pollution of the Lower Mississippi River in Louisiana.*

50. The analysis employed a combination of carbon chloroform extraction to fractionate compounds and gas chromatography to further identify compounds.

51. USEPA, *Industrial Pollution of the Lower Mississippi River in Louisiana.* For SCC evaluations, see Louisiana Stream Control Commission, *Proceedings of Meeting* (Baton Rouge, 1958–66).

52. USEPA, *Industrial Pollution of the Lower Mississippi River in Louisiana,* 6–7.

53. James F. Coerver, "New Orleans Study," unpublished paper presented to the Conference of the State Sanitary Engineers, Austin, Texas, May 18, 1976, 4–5.

54. Joel B. Goldsteen, *Danger All Around: Waste Storage Crisis on the Texas and Louisiana Gulf Coast* (Austin: University of Texas Press, 1993), 142–51.

55. Friloux, "Petrochemical Wastes as a Water Pollution Problem in the Lower Mississippi River."

56. Ralph Nader, "Statement," *Hearing Before the Subcommittee on the Environment of the U.S. Senate Committee on Commerce,* 93rd Cong., 1st sess., May 31, 1973, 88–104.

57. Robert H. Harris and Edward M. Brecher, "Is the Water Safe to Drink?" in three parts, *Consumer Reports* 39 (June 1974): 436–42; (July 1974): 538–42; and (Aug. 1974): 623–27. The claim that New Orleans residents suffered bladder and urinary cancer at an elevated rate stemmed from a 1958 USPHS Report that did not claim an environmental source for the illness. The study's authors puzzled over the fact that urinary tract cancer was higher among males and nonwhite females. It also noted that Louisiana had a high rate of lung cancer—not explained by drinking water. See Harold Dorn and Sidney Cutler, *Morbidity from Cancer in the United States,* Public Health Monograph 56 (Washington, D.C.: USPHS, 1958).

58. Harris and Brecher, "Is the Water Safe to Drink?" *Consumer Reports* 39 (June 1974): 437.

59. "Drinking Water Poses No Imminent Hazard—EPA," *New Orleans Times-Picayune,* July 20, 1974, 1.

60. "No Detectable Threshold for Chemical Carcinogen," *New Orleans Times-Picayune,* July 4, 1974, 26, and "Bottle Water Sales Flow," *New Orleans Times-Picayune,* Nov. 9, 1974, 4.

61. "Treat Water, End 50 Orleans Cancer Deaths Yearly," *New Orleans Times-Picayune,* Nov. 8, 1974, 1; and T. A. DeRouen and J. E. Diem, "New Orleans Drinking Water Controversy: A Statistical Perspective," *American Journal of Public Health* 65, no. 10 (1975): 1060–62.

62. Jean Marx, "Drinking Water: Another Source of Carcinogens?" *Science* 186 (Nov. 1974): 809–11.

63. Betty Dowty, D. R. Carlisle, and J. L. Laseter, "New Orleans Drinking Water Sources Tested by Gas Chromatography—Mass Spectrometry," *Environmental Science and Technology* 9, no. 8 (Aug. 1975): 762–65.

64. Robert W. Miller, Chief, Epidemiology Branch, National Cancer Insitute, memo, Dec. 18, 1974, in the historical files of the Louisiana Department of Health, New Orleans.

65. Robert E. Tarone and John J. Gart, "Review of the Implications of Cancer-Causing Substances in Mississippi River Water," by Robert H. Harris, unpublished paper produced by the Applied Mathematics Section of the National Cancer Institute, Jan. 10, 1975, in the historical files of the Louisiana Department of Health.

66. Talbot Page, Robert Harris, and Samuel Epstein, "Drinking Water and Cancer Mortality in Louisiana," *Science* 193 (July 1976): 55–57.

67. James F. Coerver, "New Orleans Study," unpublished paper presented to the Conference of the State Sanitary Engineers, Austin, Texas, May 18, 1976.

Chapter 10. Baton Rouge

Note: I am grateful for the assistance provided by Nan Burby and Sandra McMillan, who assembled census data and gathered much of the newspaper clipping data in Baton Rouge. I am also grateful for the helpful comments on an earlier draft of the chapter provided by Eugenie Birch of the University of Pennsylvania, Arnold Hirsch of the University of New Orleans, and Craig Colten of Louisiana State University.

1. Andrew Hurley, "Creating Ecological Wastelands: Oil Pollution in New York City, 1870–1900," *Journal of Urban History* 20, no. 3 (May 1994): 340.

2. Evidence of the problems of life next door to a refinery or chemical manufacturer were gathered in two recent investigations of environmental justice in Louisiana. See Louisiana Advisory Committee to the U.S. Commission on Civil Rights, *The Battle for Environmental Justice in Louisiana . . . Government, Industry, and the People* (Baton Rouge: The Commission, 1993). The second effort was mandated by the 1993 session of the Louisiana legislature, which called for a series of public hearings to investigate charges of environmental injustice related to the siting of petrochemical plants in the state. See Environmental Justice Group, Louisiana Department of Environmental Quality, *Final Report to the Louisiana Legislature on Environmental Justice, As Mandated by Act 767 of the 1993 Legislative Session* (Baton Rouge: Department of Environmental Quality, State of Louisiana, 1994). Also see the discussion of pollution in Baton Rouge and its effects on neighborhoods in Jim Schwab, *Deeper Shades of Green: The Rise of Blue-Collar and Minority Environmentalism in America* (San Francisco: Sierra Club Books, 1994).

3. Raymond J. Burby, "Heavy Industry, People, and Planners: New Insights on an Old Issue," *Journal of Planning Education and Research* 19 (1999): 15–25. Other studies of industrial pollution in Baton Rouge and its disparate impacts include Francis O. Adeola, "Environmental Hazards, Health, and Racial Inequality in Hazardous Waste Distribution," *Environment and Behavior* 26 (1994): 99–126 and Adeola, "Demographic and Socioeconomic Differentials in Residential Propinquity to Hazardous Waste Sites and Environmental Illness," *Journal of the Community Development Society* 26 (1995): 15–39.

4. Historical treatments of urban industrial pollution are provided in William Cronon, *Nature's Metropolis: Chicago and the Great West* (New York: W. W. Norton, 1991); Hurley, "Creating Ecological Wastelands"; Andrew Hurley, "Busby's Stink Boat and the Regulation of Nuisance Trades, 1865–1918," in *Common Fields: An Environmental History of St. Louis,* ed. Andrew Hurley (St. Louis: Missouri Historical Society Press, 1997), 145–62; Christine Meisner Rosen, "Noisome, Noxious, and Offensive Vapors, Fumes and Stenches in American Towns and Cities, 1840–1865," *Historical Geography* 25 (1997): 49–82; and Joel A. Tarr, *The Search for the Ultimate Sink: Urban Pollution in Historical Perspective* (Akron, Ohio: University of Akron Press, 1996).

5. Common law provides that landowners may not use their property in ways that impair the rights of other people. That is, they cannot create nuisances. Two types of nuisance suits were common in the nineteenth century: private and public. Private nuisance suits are used to deal with situations where an individual property owner is harmed, such as the pollution of someone's well by adjacent industrial activity.

Public nuisances involve activities that adversely affect the public in general. Activities that create air and water pollution over a large area are examples of public nuisances. Susan J. Buck, *Understanding Environmental Administration and Law* (Washington, D.C.: Island Press, 1991), 61–63.

6. The inability of common law and nuisance litigation to deal with industrial pollution is discussed in most treatments of urban pollution in the nineteenth century, e.g., Hurley, "Busby's Stink Boat"; Rosen, "Noisome, Noxious, and Offensive Vapors"; and David Stradling, "Civilized Air: Coal, Smoke, and Environmentalism in America, 1880–1920," Ph.D. dissertation, University of Wisconsin, Madison, 1996. Also see Denis J. Brion, *Essential Industry and the NIMBY Phenomenon* (New York: Quorum Books, 1991); Martin V. Melosi, "Hazardous Waste and Environmental Liability in Historical Perspective," *Houston Law Review* 25, no. 4 (July 1988): 741–80; and Stanley K. Schultz, *Constructing Urban Culture: American Cities and City Planning, 1800–1920* (Philadelphia: Temple University Press, 1989), 42–80. Contemporary problems in the use of nuisance suits are discussed by sociologist Stela Capek, who notes, "In many contaminated communities getting satisfactory compensation is the most difficult struggle of all. The major social actors involved in compensation are the polluting corporations themselves and the federal government. Corporations generally do not acknowledge fault, and if pressed, prefer to engage in out-of-court settlements." Stela M. Capek, "The Environmental Justice Frame: A Conceptual Discussion and an Application," *Social Problems,* 40, no. 1 (Feb. 1993): 14. Greater use of compensatory approaches, however, is argued for in Thomas Lambert and Christopher Boerner, "Environmental Inequity: Economic Causes, Economic Solutions," *Yale Journal of Regulation* 14, no. 1 (1997): 195–234.

7. Melosi ("Hazardous Waste and Environmental Liability," 769) notes, "By the 1880s many cities passed statutes restricting noxious manufacturers to the outskirts." Thirty years later, this had become fairly standard practice, as noted in George M. Price, *The Modern Factory: Safety, Sanitation, and Welfare* (New York: John Wiley, 1914). In this early manual on industrial sanitation, Price notes, "Where excessive noise becomes a public nuisance, municipal regulations require plants causing it to be located outside of the city limits and at a distance from human habitation" (283).

8. Rosen, "Noisome, Noxious, and Offensive Vapors," 51–52.

9. This seems to be particularly true of chemical plants. Craig Colten and Peter Skinner write, "Selecting remote manufacturing locations enabled chemical producers to guard against recognized liabilities—such as off-site property damage risks due to explosions, direct contact by children or unsuspecting adults, or water and air pollution—and acknowledged an endangerment potential." Craig E. Colten and Peter Skinner, *The Road to Love Canal: Managing Industrial Waste before EPA* (Austin: University of Texas Press, 1996), 47.

10. Craig E. Colten, "Historical Hazards: The Geography of Relict Industrial Wastes," *Professional Geographer* 42, no. 2 (1990): 143–56.

11. Martin V. Melosi, introduction to *Pollution and Reform in American Cities, 1870–1930,* ed. Martin V. Melosi (Austin: University of Texas Press, 1980), 8.

12. Benjamin Marsh, *An Introduction to City Planning: Democracy's Challenge to the*

American City (New York: B. C. Marsh, 1909), and John Nolen, *City Planning* (New York: D. Appleton, 1916). Nolen's plan for the industrial city of Kingsport, Tennessee, is often cited as an example of how early U.S. planners thought industry should be accommodated.

13. American Public Health Association, Committee on the Hygiene of Housing, *Planning the Neighborhood: Standards for Healthier Housing* (Chicago: Public Administration Service, 1948).

14. Institute for Training in Municipal Administration, *Local Planning Administration,* 2nd ed. (Chicago: International City Managers Association, 1948); Theodore Pasma, *Organized Industrial Districts* (Washington, D.C.: USGPO, June 1954).

15. The municipal smoke-control movement in England is described in Carlos Flick, "The Movement for Smoke Abatement in Nineteenth-Century Britain," *Technology and Culture* 21 (Jan. 1980): 29–50. Smoke-control efforts in the U.S. are described in Tarr, *The Search for the Ultimate Sink,* and Stradling, "Civilized Air."

16. American Society of Planning Officials, "Performance Standards for Industrial Zoning," Information Report No. 32, Chicago, Nov. 1951, 1–3.

17. Dorothy A. Muncy, "Land for Industry—A Neglected Problem," *Harvard Business Review* 32 (March–April 1954): 61.

18. David M. Welborn, "Conjoint Federalism and Environmental Regulation in the United States," *Publius: The Journal of Federalism* 18 (Winter 1988): 27–43.

19. The oil industry in Louisiana and the founding of Standard Oil's petrochemical business in Baton Rouge are chronicled in the following: Hodding Carter, *John Law Wasn't So Wrong* (Baton Rouge: Esso Standard Oil Company, 1952); and Kenny A. Franks and Paul F. Lambert, *Early Louisiana and Arkansas Oil: A Photographic History, 1901–1946* (College Station: Texas A&M University Press, 1982).

20. Quoted in *Baton Rouge Record,* Oct. 18, 1916.

21. Over the years, the name of the refinery changed with that of its corporate parent company: Standard Oil (of Louisiana, a subsidiary of Standard Oil), Jersey Standard (after the breakup of the Standard Oil trust), Esso Standard Oil, Humble Oil, and Exxon. In this chapter, I use the modern name, Exxon, to refer to the refinery and immediately surrounding industrial complex after the breakup of the Standard Oil trust.

22. "50th Anniversary Issue," *Esso News* 36 (1959).

23. George S. Gibb and Evelyn H. Knowlton, *The History of Standard Oil Company (New Jersey): The Resurgent Years, 1911–27* (New York: Harper, 1956), 560.

24. These factors were listed in a survey by *Chemical Week,* Dec. 7, 1957. Also see Robert Nance McMichael, "Plant Location Factors in the Petrochemical Industry in Louisiana," Ph.D. dissertation, Louisiana State University, Baton Rouge, 1961.

25. Andrew Hurley, *Environmental Inequalities: Class, Race, and Industrial Pollution in Gary, Indiana, 1945–1980* (Chapel Hill: University of North Carolina Press, 1995), 38.

26. Robert Fisher, "The Urban Sunbelt in Comparative Perspective: Houston in Context," in *Essays on Sunbelt Cities and Recent Urban America,* ed. Raymond A. Mohl et al., eds. (College Station: Texas A&M University Press, 1990), 34. Throughout this pe-

riod, Baton Rouge looked to Houston as a model of the modern industrial city, and a delegation from the local chamber of commerce visited there in the 1950s to pick up tips on industrial promotion. As has been recounted numerous times, Houston, among all U.S. cities, stands out for the degree to which it has embraced laissez-faire capitalism. See, for example, Barry Kaplan, "Houston: The Golden Buckle of the Sunbelt," in *Sunbelt Cities: Politics and Growth since World War II*, ed. Richard M. Bernard and Bradley R. Rice (Austin: University of Texas Press, 1983), 199.

27. City of Baton Rouge and Parish of East Baton Rouge, *Conservation and Environmental Resource Element, Comprehensive Land Use and Development Plan* (Baton Rouge: City of Baton Rouge and Parish of East Baton Rouge, July 1992), 22.

28. Parish of East Baton Rouge and City of Baton Rouge, *The Plan of Government of the Parish of East Baton Rouge and the City of Baton Rouge*, adopted Aug. 12, 1947, effective Jan. 1, 1949 (Baton Rouge: City of Baton Rouge and Parish of East Baton Rouge, 1948).

29. This did not come to light publicly until 1970, when the city was debating funds to upgrade the landfill to modern standards. The landfill superintendent, seeking support for his budget request, noted that trucks from the petrochemical industry ran bumper to bumper to the landfill (*Baton Rouge Morning Advocate*, 1969).

30. Schwab, *Deeper Shades of Green*, 249. According to Schwab, a planner and writer working for the American Planning Association, "Together, the oil-refining and chemical industries soaked up some $1.1 billion in taxpayer largesse during the 1980s." Also see Paul Templet, "The Full Economic Costs of Louisiana Oil/Gas and Petrochemical Industries," People First: Developing Sustainable Communities and Institute of Environmental Studies, Louisiana State University, Baton Rouge.

31. Also, escape from the expansion of black neighborhoods into formerly white working-class neighborhoods adjacent to the north Baton Rouge industrial complex has been a factor in the exodus. Livingston Parish residents surveyed by Shanafelt in the mid-1970s listed that as a reason for leaving the city. See Raymond Edgar Shanafelt, "The Baton Rouge-New Orleans Petrochemical Industrial Region: A Functional Region Study," Ph.D. dissertation, Louisiana State University, Baton Rouge, 1977. In the 1990s the fastest-growing parishes in Louisiana were adjacent to East Baton Rouge Parish, which suggests the flight of residents continued.

32. Harland Bartholomew and Associates, *The Master City-Parish Plan for Metropolitan Baton Rouge, Louisiana, Preliminary Report Upon Chapter VIII: Land Use* (St. Louis: Harland Bartholomew and Associates, May 1946), 2.

33. City-Parish Planning Commission, *Land Use and Zoning Study* (Baton Rouge: City-Parish Planning Commission, July 1958), 4.

34. *Baton Rouge Morning Advocate*, Jan. 11, 1956, and July 7, 1957.

35. In 1945 the eight oil refineries in Louisiana formed the Louisiana Refiners Waste Control Council to find common solutions to pollution problems. Their major focus, however, was reduction of waste emissions to streams rather than air pollution. See *Esso News* 33 (Feb. 13, 1956): 1. Over the succeeding years and continuing to this present day, the industry's response to public concerns about pollution has been to

form committees or task forces to "study the problem." This reflects a general prefer-
ence of the oil industry to deal with problems internally rather than have solutions
imposed by government regulators. See Martin V. Melosi, *Coping with Abundance:
Energy and Environment in Industrial America* (Philadelphia: Temple University Press,
1985), 267.

36. *Baton Rouge State Times*, Oct. 27, 1965.

37. *Baton Rouge Morning Advocate*, Sept. 11 and 14, 1965; *Winston-Salem Journal*, Sept.
13, 1965; *Baton Rouge State Times*, Oct. 11, 1965.

38. Schwab, *Deeper Shades of Green*, 250.

39. *Baton Rouge Morning Advocate*, Jan. 13, 1990.

40. Schwab, *Deeper Shades of Green*, 259.

41. Urbanist Jane Grant notes that residents living near heavy industry understand pollution in much more personal terms than white, middle-class Americans,
who are not reminded daily of their vulnerability to hazards by odors, dust, and noise
from nearby industrial operations. For the poor, exposure becomes not just a threat to
health but also, according to Grant, a threat to "psychological and family well-being,
community stability, and economic viability." Jane A. Grant, "Assessing and Managing
Risk in the Public Sector: An Urban Hazardous Waste Landfill," *Journal of Urban
Affairs* 16 (1996): 338.

42. The inability of urban blacks to be effective in righting various grievances
during the first half of the twentieth century is discussed in David R. Goldfield, "Black
Political Power and Public Policy in the Urban South," in *Urban Policy in Twentieth-Century America*, ed. Arnold R. Hirsch and Raymond A. Mohl (New Brunswick, N.J.: Rut-
gers University Press, 1993), 159–82. A civil rights suit currently in litigation evidences
the weak political position of blacks in Baton Rouge. In this suit, the U.S. Department
of Justice claims the City-Parish violated blacks' civil rights in its decision in 1982 to
abolish the Baton Rouge City Council just as blacks were gaining a voting majority in
the city.

43. Factors behind racial transition in industrial neighborhoods in other cities also
have been difficult to determine. Urban historian Andrew Hurley studied industrial
neighborhoods in St. Louis and Gary, where racial transition appeared to follow on
the heels of white flight to the suburbs, and could not attribute it to industrial hazards.
In a third case, the oil and petrochemical neighborhoods on Long Island in the nineteenth century, however, he found evidence that the growing petrochemical industry
hastened the exit of well-to-do households. Andrew Hurley, "Fiasco at Wagner Electric: Environmental Justice and Urban Geography in St. Louis," *Environmental History*
2 (Oct. 1997): 462. Also see Andrew Hurley, *Class, Race, and Industrial Pollution in Gary,
Indiana, 1945–1980* (Chapel Hill: University of North Carolina Press, 1995), and Hurley,
"Creating Ecological Wastelands," 360.

44. These efforts are recounted in Schwab, *Deeper Shades of Green*.

45. Susan L. Cutter and John Tiefenbacher, "Chemical Hazards in Urban America," *Urban Geography* 12 (Oct. 1991): 422, 425.

46. Enacted in 1986, Title III of the Superfund Amendments, the Emergency

Planning Community Right-to-Know Act (EPCRA) required companies to disclose releases of toxic substances to the ambient environment or transfer of hazardous wastes to off-site locations.

47. "River Sentinel 1995: People Who Can Be Harmed by Toxic Releases in Louisiana's Mississippi River Industrial Corridor—Baton Rouge Community," map with accompanying data, Center for Energy and Environmental Studies, Southern University, Baton Rouge, 1997.

48. Michael H. Brown, "The National Swill: Poisoning Old Man River," *Science Digest* 94 (June 1986): 56–65, 82–85.

49. Schwab, *Deeper Shades of Green,* 214.

50. Harvey L. White, "Hazardous Waste Incineration in Minority Communities," in *African-American Environmental Education Program,* prepared for Lyons, Louisiana Community Leaders (New Orleans: Gulf Coast Tenants Organization, 1993): 21–35.

51. *Baton Rouge Morning Advocate,* Feb. 26, 1958. The planners were following the dictates of best practice. In a 1954 article in the *Harvard Business Review* urban economist Dorothy Muncy argued that planners were doing too little to protect valuable industrial land from being co-opted by less important land uses, such as residential subdivisions. Muncy, "Land for Industry—A Neglected Problem," 51–63.

52. The following planning documents were examined to determine what, if anything, Baton Rouge city and parish government had considered to deal with the environmental pollution and other harms that accompany heavy industry: Harland Bartholomew and Associates, *The City-Parish Plan, Baton Rouge, Louisiana, Preliminary Report Upon Chapter One: Scope and Objectives of the Plan* (St. Louis: Harland Bartholomew and Associates, Aug. 1945); City-Parish Planning Commission, *Land Use and Zoning Study;* City and Industrial Planners, *Proposed Plan for Future Land Use in East Baton Rouge Parish* (Baton Rouge: City and Industrial Planners, Feb. 1972); City-Parish Planning Commission, *Comprehensive Development Study: A Detailed Land Use Development Plan, Sector 10* (Baton Rouge: City-Parish Planning Commission, Oct. 1973); and City of Baton Rouge and Parish of East Baton Rouge, *Horizon Plan Summary* (Baton Rouge: City of Baton Rouge and Parish of East Baton Rouge, Feb. 1991).

53. In addition to individual citizens, various groups offered testimony. These included the Coalition for Community Action, North Baton Rouge Environmental Association, Louisiana Environmental Action Network, Sierra Club Legal Defense Fund, and Southern University.

54. Environmental Justice Group, Louisiana Department of Environmental Quality, *Final Report to the Louisiana Legislature on Environmental Justice.*

55. Maury Klein and Harvey A. Kantor, *Prisoners of Progress: American Industrial Cities, 1850–1920* (New York: Macmillan, 1976), 432. In this regard it is interesting to note that immediately after issuing its report on air pollution control in 1950 the American Society of Planning Officials, in response to requests from numerous communities, issued its next report on how communities could attract additional industry by offering an array of inducements. See American Society of Planning Officials, *Community Inducements to Industry,* Information Report no. 22 (Chicago: ASPO, Jan. 1951).

Chapter 11. The Popular Geography of Illness in the Industrial Corridor

1. Eddystone C. Nebel III, *Factors Affecting the Location of the Petrochemical Industry in the Gulf South* (Baton Rouge: Louisiana State University Press 1971), 58.

2. "Louisiana Oil Prospers by 'No Politics' Rule," *Oil and Gas Journal* (Jan. 27, 1964): 114.

3. "Petrochemical Boom Hits Tiny Geismar," *Oil and Gas Journal* (May 4, 1964): 50.

4. Homesite Company, *Louisiana Mississippi River Directory of Petrochemical Industries* (Baton Rouge: Homesite Company, 1995).

5. See Mary Ann Sternberg, *Along the River* (Baton Rouge: Louisiana State University Press, 1996), for a complete account, mile by mile, of plantations that have been sold to industry. See also Leon Bubber Geismar, interview by author, Gonzales, La., May 22, 1997. Geismar, a 71-year-old primary landowner in the area listed, to the best of his memory, all of the industrial sites and the wealthy white families they had purchased the sites from. He had also been a commercial real estate agent during this period and facilitated many of the transactions.

6. For chemical production data see Loren C. Scott, *The Chemical Industry in Louisiana* (Baton Rouge: Louisiana Chemical Association, 1996). For hazardous waste production see Pat Costner and Joe Thornton, *We All Live Downstream: The Mississippi River and the National Toxics Crisis* (Seattle: Greenpeace, 1989), 91–92.

7. Costner and Thornton, *We All Live Downstream,* 93. Waste that is produced as a result of oil and gas production is exempt from EPA hazardous waste regulations. This waste can contain heavy metals and other toxic materials and is often dumped in open landfills in south Louisiana.

8. Phil Brown, a leading scholar of popular epidemiology, is cautious about the term *epidemiology* as the popular version includes not only data but concern for the social structures and power contexts in which the activity occurs. He retains the word *epidemiology* in the concept of popular epidemiology because the starting point is the search for rates and causes of diseases. For a further discussion see Phil Brown, "Popular Epidemiology: Community Response to Toxic Waste-Induced Disease in Woburn, Massachusetts," *Science, Technology, and Human Values* 12, nos. 3 and 4 (1987): 78–85.

9. At the time the Rollins hazardous waste landfill was the fourth largest in the nation and served as a disposal site for major chemical companies such as Exxon and BASF. See Jim Schwab, *Deeper Shades of Green* (San Francisco: Sierra Club Books, 1994), 218.

10. Florence Robinson, telephone interview by the author, Alsen, La., July 31, 1997. She noted in the interview that her son's headache problems ceased when he moved away.

11. This material is taken from a report by Florence Robinson submitted as an appendix as part of the public record in the hearing transcript for the Louisiana Department of Environmental Quality (LDEQ), *Environmental Justice Public Hearing* (Baton Rouge: LDEQ, March 19, 1994).

12. Kay and Chris Gaudet, interview by author, St. Gabriel, La., June 6, 1997.

13. Kay Gaudet's observations were made public in an editorial by Ellie Hebert which appeared in the *Iberville Post South*, Sept. 17, 1987, A-2.

14. Schwab, *Deeper Shades of Green*, 237.

15. From "St. Gabriel Awaits Study on Miscarriages," *New Orleans Times-Picayune*, Sept. 10, 1989, sec. B, p. 9.

16. From LuAnn E. White, Frances J. Mather, and Jacquelyn R. Clarkson, *Final Report: St. Gabriel Miscarriage Investigation* (New Orleans: Tulane University, School of Public Health and Tropical Medicine, Sept. 27, 1989).

17. White, *Final Report*, vi.

18. Richard Clapp's opinions were stated in an article by James O'Byrne, "Study of St. Gabriel Miscarriages Flawed," *New Orleans Times-Picayune*, May 16, 1990, B-10.

19. Paul Templet was secretary of the DEQ for four years. As an environmental scientist he was a staunch advocate for citizen participation and stringent environmental enforcement. His administration was involved in reviewing the report that Tulane submitted. He admitted to me in his interview (July 18, 1997) his disappointment in the study which the state funded.

20. Templet interview. In my interview with the Gaudets, Kay mentioned that some of the black women were reluctant to talk about their miscarriages because someone in their family worked for one of the plants. From my own observation, there is a general mistrust of government-sponsored surveys and programs in this area that tends to be more pronounced in the minority populations, probably due to past experiences.

21. Templet interview.

22. White, *Final Report*, v.

23. Louisiana Advisory Committee to the U.S. Commission on Civil Rights, *The Battle for Environmental Justice in Louisiana* (Baton Rouge: Louisiana Advisory Committee to the U.S. Commission on Civil Rights, Sept. 1993), 37.

24. Louisiana Tumor Registry, Office of Public Health and Hospitals, *Executive Summary, Cancer Incidence in South Louisiana, 1983–1986* (New Orleans: LSU Medical Center, July 1991), 1.

25. David Ozonoff and Leslie I. Boden, "Truth and Consequences: Health Agency Responses to Environmental Health Problems," *Science, Technology, and Human Values* 12, nos. 3 and 4 (1987): 70–77.

26. According to Ozonoff and Boden (ibid., 74) there can be an enormous increase in disease rates that are ruled not statistically significant because the population sample is not large enough. Using epidemiological methods, they cite the example of a 30 percent increase in the rate of leukemia in females living in the five towns near the Plymouth nuclear plant being declared not statistically significant.

27. Richard A. Couto, cited in Brown, "Popular Epidemiology," 82.

28. Robinson interview.

29. The Louisiana Chemical Association is both a lobbying and public relations organization providing services to the chemical industry throughout the state.

30. Louisiana Advisory Committee to the U.S. Commission on Civil Rights, *The Battle for Environmental Justice in Louisiana*, 37–38.

31. Louisiana Tumor Registry, Office of Public Health and Hospitals, *Executive Summary*, 2.

32. From Patricia Andrews, Catherine Correa, Vivien Chen et al., *Cancer in Louisiana* (Baton Rouge: Louisiana Tumor Registry, 1983), 10, 137. For white and black men the incidence rate of lung cancer is 36 percent and 15 percent above the national averages respectively. For white and black women it is 10 percent above and 12 percent below the national norms.

33. Louisiana Tumor Registry, Office of Public Health and Hospitals, *Executive Summary*, 2.

34. Ibid.

35. "Poor" is defined as a family income of less that $10,000 per year and the mortality rate of this groups is 3.22 times that of those making over $30,000 per year. For the low-income group making between $10,000 and $30,000 per year the mortality rate is 2.34 times that of the higher income group. For a complete summary of the research see Paula M. Lantz, James S. House et al., "Socioeconomic Factors, Health Behavior, and Mortality," *Journal of the American Medical Association* 279, no. 21 (June 3, 1998): 1703.

36. The study only included people over the age of 25 and had an overrepresentation of blacks in the sample. The researchers also found when studying the alcohol consumption habits of the various income groups that lower income groups actually consumed less alcohol that their higher income counterparts. From Lantz et al., "Socioeconomic Factors," 1706.

37. Lantz, "Socioeconomic Factors," 1707

38. Ibid.

39. Paul Templet, personal correspondence with author, n.d.

40. Albertha Hasten, interview by author, Whitecastle, La., June 24, 1997.

41. From Nancy Krieger and Mary Bassett, "The Health of Black Folk: Disease, Class, and Ideology in Science," *Monthly Review* 38, no. 3 (July–Aug. 1986): 81–82.

42. Both my ethnographic fieldwork in the area and public hearing transcripts from permitting hearings verify that many of the residents have lived in the area for generations, well before the advent of the chemical industry in the 1960s and 1970s. One comprehensive study by Raymond Burby, *Through Their Eyes: Survey Results of Lower Income Residents in the Louisiana Industrial Corridor* (New Orleans: UNO Research Center, 1995), found that among low-income residents, 78 percent owned their own home, 30 percent have lived in the same house 16 to 30 years and another 30 percent have lived in the same house 31 or more years. Because of the large number of homeowners in the region, I surmise that many people who have moved more recently probably did so to live with family members or build a home on family property. Furthermore, according to Loren Schweninger in *Black Property Owners in the South, 1790–1915* (Urbana: University of Illinois Press, 1990), over half of the black rural landowners in the eight states composing the lower South were in Louisiana. Furthermore, their numbers were concentrated along the river parishes (80, 120).

43. Adeline Gordon Levine, *Love Canal: Science, Politics, and People* (Lexington, Mass.: Lexington Books, 1982), 102.

44. Richard Clapp, letter to author, Aug. 5, 1998.

45. Institute for Environmental Issues and Political Assessment, *River Sentinel: Environmental, Health, and Social Trends in the Lower Mississippi River Corridor* (Baton Rouge: Center for Energy and Environmental Studies, Southern University, 1994), 35–36.

46. Vivien Chen, telephone interview by author, Oct. 8, 1998.

47. Ibid.; Clapp letter.

48. From an article by Vivien Chen et al., titled "Cancer Incidence in the Industrial Corridor: An Update," *Journal of the Louisiana State Medical Society* 150, no. 4 (April 1998): 158–67. This article, which was part of the special issue devoted to cancer in Louisiana, concluded with a call for cancer control programs that emphasize the prevention and cessation of tobacco use (166).

49. Burby, *Through Their Eyes*, 2.

50. Patricia Williams, interview by author, Lafayette, La., Oct. 8, 1998.

51. Ed Bradley did a CBS special on the Grand Bois community titled "On Assignment: Town Under Siege," on Dec. 23, 1997. He also publicized the town's plight on "The Charlie Rose Show" on Dec. 22, 1997. Oilfield waste is largely exempt from the EPA regulations as it is automatically classified nonhazardous and the companies are not required to disclose the content of the waste material.

52. Williams interview.

53. Ibid.

54. Williams, unpublished manuscript, "Critique of the Louisiana Tumor Registry," n.d.

55. Williams interview.

56. Ibid.

57. Williams, personal correspondence with author, n.d.

58. John H. Cushman, "Rising Children's Cancer Rates Reshapes U.S. Research Efforts," *New Orleans Times-Picayune*, Sept. 29, 1997, A-1, A-8. Statistics suggest that a newborn child has a 1 in 600 chance of contracting cancer by the age of 10.

59. Ibid.

60. Laurie Smith Anderson, "Doctor Says Cluster of Cancer Cases Merits Investigation," *Baton Rouge Advocate*, March 31, 1996, H-2, H-8.

61. Ibid.

62. The M.D. Anderson cancer facility in Houston is actually closer than St. Jude. Another of Williams's worries is that the cases that are diagnosed and referred out of state might not be recorded properly as Louisiana cases.

63. Vivien Chen et al., *Cancer Incidence in Louisiana, 1988–92* (New Orleans: Louisiana Tumor Registry, 1996), 5, as cited in Williams's manuscript, "Critique of the Louisiana Tumor Registry," 2.

64. Williams, personal correspondence, n.d.

65. Charles Flanagan, "Shintech Hearing Transcript," Louisiana Department of Environmental Quality, Convent, La., Jan 24, 1998, 343–45.

66. Ozonoff and Boden, "Truth and Consequences," 72.

Chapter 12. Fish Diversity in a Heavily Industrialized Stretch
of the Lower Mississippi River

Note: The author wishes to thank Royal D. Suttkus for his many years of dedication to surveying the fishes of streams in the Gulf South region, including the lower Mississippi River. Without this effort, this chapter would not have been possible. Special thanks are also extended to Michael S. Taylor for help with computer graphics for this chapter and the presentation given in the Centuries of Change Symposium.

1. Louisiana Department of Environmental Quality, Early Warning Organic Compound Detection System (EWOCDS), http://www.deq.state.la.us/owr/ewtxt/htm.

2. N. N. Rabalais, "Mississippi River Water Quality: Status, Trends, and Implications," in *Proceedings of the 1996 Louisiana Environmental State of the State Conference,* ed. Lida Durrant (Baton Rouge: Environmental Research Consortium of Louisiana, 1996).

3. J. A. Baker, K. J. Killgore, and R. L. Kasul, "Aquatic Habitats and Fish Communities in the Lower Mississippi River," *Reviews in Aquatic Sciences* 3, no. 4 (1991): 313–56.

4. Tristram Kidder, "The Role of Native Americans in Shaping the Environment and Geography of New Orleans," in this volume.

5. Rabalais, "Mississippi River Water Quality."

6. T. Coburn and C. Clement, eds., *Chemically Induced Alterations in Sexual and Functional Development: The Wildlife-Human Connection,* Advances in Modern Environmental Toxicology 21 (Princeton: Princeton Scientific Publishing, 1992).

7. C. F. Bryan, Allen Rutherford, and B. Walker-Bryan, "Acidification of the Lower Mississippi River," *Transactions American Fisheries Society* 121 (1992): 369–77.

8. Rabalais, "Mississippi River Water Quality."

9. M. D. Henrich, A. E. Hindrichs, and M. B. Fleming, *Mississippi River Toxics Inventory Project: Fish and Shellfish Sampling Plan* (Baton Rouge: Louisiana Department of Environmental Quality, 1995), 29.

10. Henry L. Bart Jr., P. Martinat, A. Abdelghani, and S. L. Taylor, "Fish Population and Community Responses to Aquatic Contamination in Swamps of the Lower Mississippi River System," *Ecotoxicology* 7 (1998): 325–34.

11. J. P. Sumpter and S. Jobling, "Vitellogenesis as a Biomarker for Estrogenic Contamination of the Aquatic Environment," *Environmental Health Perspectives* 103 (1995): 173–78; L. C. Folmar, N. D. Denslow, V. Rao, M. Chow, and D. A. Crain, "Vitellogenin Induction and Reduced Serum Testosterone Concentrations in Feral Male Carp (Cyprinus carpio) Captured Near a Major Metropolitan Sewage Treatment Plant," *Environmental Health Perspectives* 104 (1996): 1096–1101.

12. J. V. Conner and C. Fred Bryan, "Review and Discussion of Biological Investigations in the Lower Mississippi River and Atchafalaya River," *Proceedings of the 28th Annual Conference of the Southeastern Association of Game and Fish Commission* 28 (1974): 429–41.

13. D. S. Jordan, "List of Fishes Collected in the Vicinity of New Orleans by R. W. Shufeldt, U.S.A.," *Proceedings U.S. National Museum* 7 (1884): 318–22.

14. J. R. Kelly Jr., "A Taxonomic Survey of the Fishes of the Delta National

Wildlife Refuge, with Emphasis on Distribution and Abundance," M.S. thesis, Louisiana State University, 1965.

15. V. A. Guillory, "Distribution and Abundances of Fish in Thompson Creek and the Lower Mississippi River, Louisiana," M.S. thesis, Louisiana State University, 1974; and V. A. Guillory, "Fishes of the Lower Mississippi River near St. Francisville, Lousiana," *Louisiana Academy of Sciences* 45 (1982): 108–21.

16. C. R. Robbins, R. M. Bailey, C. E. Bond, J. R. Brooker, E. A. Lachner, R. N. Lea, and W. B. Scott, *Common and Scientific Names of Fishes From the United States and Canada,* 5th ed. (Bethesda, Md.: American Fisheries Society, Special Publication, 1991), 183.

17. Not taken in the study area, but F. A. Cook in *Freshwater Fishes of Mississippi* (Jackson: Mississippi Game and Fish Commission, 1959), reported on specimens removed from fish caught in the vicinity of Vicksburg, Mississippi.

18. J. A. Thomerson, "The Bull Shark, Carcharhinus leucas, from the Upper Mississippi River near Alton, Illinois," *Copeia* (1977): 166–68.

19. Henrich, Hindrichs, and Fleming, *Mississippi River Toxics Inventory Project.*

20. Folmar et al., "Vitellogenin"; and J. E. Harries, D. A. Sheahan, S. Jobling, P. Matthiessen, and P. Neall, "Estrogenic Activity in Five United Kingdom Rivers Detected by Measurement of Vitellogenesis in Caged Male Trout," *Environmental Toxicology and Chemistry* 16 (1997): 534–42.

Contributors

Barbara Allen is Director of the Graduate Program in Science and Technology at Virginia Tech's Northern Virginia campus in the D.C. area. In 1999, she earned her Ph.D. at Rensselaer Polytechnic Institute in Science and Technology Studies, where she researched the strategies of citizen-expert alliances toward initiating positive environmental change in Louisiana's chemical corridor. She has published other articles on this subject in various volumes including *Technologies of Landscape, Michigan Feminist Review,* and *Contemporary Justice Review.*

Henry L. "Hank" Bart is Associate Professor of Ecology, Evolution, and Organismal Biology at Tulane University and Director and Curator of Fishes of the Tulane Museum of Natural History. He is editor of *Tulane Studies in Zoology and Botany* and *Occasional Papers Tulane University Museum of Natural History.* A major emphasis of his research is biotic changes in large rivers, including the lower Mississippi River. A native of New Orleans, Bart earned his Ph.D. in zoology (1985) from the University of Oklahoma.

Raymond J. Burby is Professor of City and Regional Planning at the University of North Carolina at Chapel Hill. He is former coeditor of the *Journal of the American Planning Association* and author of numerous books and articles dealing with land use and environmental problems, including *Cooperating with Nature* (Joseph Henry / National Academy Press, 1998) and *Making Governments Plan* (Johns Hopkins University Press, 1997). Burby's interest in the lower Mississippi valley developed while he served as DeBlois Chair and Distinguished Professor at the University of New Orleans between 1992 and 2000.

Craig E. Colten is Associate Professor of Geography at Louisiana State University. A native of Louisiana, he has investigated human transformation of Mississippi River tributaries from the French Broad to the Illinois and now is focusing on environmental change in the greater New Orleans area. He is coeditor of *Historical Geography* and coauthor of *The Road to Love Canal.*

Donald Davis is Research Professor in the Energy Program at Louisiana State University and Administrator of the Louisiana Applied and Educational Oil Spill Research and Development Program. His professional career has focused on investigating various human/land issues in Louisiana's wetlands. In this regard, he has for more than thirty years had an interest in the Mississippi River's direct and indirect impact on the people and resources within Louisiana's alluvial wetlands.

Gay M. Gomez, a native of New Orleans, earned her Ph.D. in geography at the University of Texas. Currently she is Assistant Professor of Geography at McNeese State University in Lake Charles, Louisiana. She is the author of *A Wetland Biography: Seasons on Louisiana's Chenier Plain.*

Ari Kelman teaches urban and environmental history at the University of Denver. His book, *A River and Its City: An Environmental History of New Orleans,* is forthcoming from the University of California Press. Dr. Kelman's interest in the lower Mississippi Valley stems from his fascination with the relationship between built and natural environments.

Tristram R. Kidder is Associate Professor of Anthropology at Tulane University. He has published papers on premodern human impacts on the environment in *Advances in Historical Ecology* and numerous articles about the prehistory and geomorphology of the Mississippi River and delta. His interest in the region stems from undergraduate archaeological fieldwork in southeast Louisiana and his work with the Lower Mississippi Survey while a graduate student at Harvard University.

Christopher Morris is Associate Professor of History at the University of Texas at Arlington. He is author of *Becoming Southern: The Evolution of a Way of Life, Warren County and Vicksburg, Mississippi, 1770-1860,* and is writing a book on the environmental and social history of the lower Mississippi Valley, 1500 to the present.

George S. Pabis is Assistant Professor of History at Georgia Perimeter College. His research interests focus on the attempt of engineers to understand and mitigate natural environmental changes. Dr. Pabis's interests in America's rivers began as a young boy when he went swimming in the Ohio River and noticed that it looked more like a lake than anything else. His future work examines the methods scientists have used to comprehend the role of the environmental catastrophe in the extinction of the dinosaurs.

Todd Shallat teaches writing and history at Boise State University. A Ph.D. from Carnegie-Mellon University, he specializes in the history of rivers and environmental engineering. His books include *Structures in the Stream* (winner of the Henry Adams Prize for the best history of the federal government) and a forthcoming study of the Corps of Engineers on the Mississippi River.

Index